Elementary Particle Physics for Enthusiasts

Yoshiki Teramoto

ISBN: 1512142948

ISBN-13: 978-1512142945

Preface

This book is intended for the readers who like to understand the elementary particle physics, especially the Higgs particles, the classification and the dynamics of quarks and the particle-antiparticle symmetry violation, in addition to the other topics in the standard model of the elementary particle physics. In this book, most of the concepts in the particle physics are explained "directly" by figures and words, but without using equations. This "directly" means that analogies are avoided as much as possible. For example, typical books describe the Higgs mechanism as the drag forces by the Higgs particles. This book, however, explains it using the Heisenberg uncertainty principle, which is interpreted as a consequence of the oscillating nature of elementary particles. The fundamental concept underlying in this book is "oscillation" in the microscopic world.

The target readers are those who have strong interests in the particle physics and who already have some knowledge on this subject. The knowledge of high-school physics and mathematics is assumed. I avoid using equations as much as possible in this book.

Contents

Chapter 1

Orientation

We introduce some basic words and concepts in this chapter. The reader who is familiar with such introductory concepts may want to skip this chapter.

1.1   What are elementary particles?

What makes matters in everyday life? Water in a river is made of water molecule, H_2O, for example. A water molecule is made of two hydrogen atoms and one oxygen atom. A hydrogen atom is made of a hydrogen nucleus and an **electron** (e) . The hydrogen nucleus is made of a proton (p). And the proton is made of **quarks**. Since we don't know the substructures of electrons and quarks, we call them **elementary particles**. The elementary particles are the building blocks of all the matters in everyday life.

All the matters are made of elementary particles and vacuum. You may imagine this as dusts, i.e. elementary particles, are floating in the air, i.e. vacuum.

In addition to electrons and quarks, there are about twenty kinds of elementary particles known today. One of the most familiar elementary particles is **photon**. Though light is considered as a kind of electromagnetic wave, it is also a kind of elementary particles. We call this feature as **wave-particle duality**.

1.2   Dark matter and dark energy

Some phenomena, observed in the universe, can not be explained by known elementary particles only. You may have heard of **Dark Matter** and **Dark Energy**. Those may be necessary to explain the observed cosmological phenomena. In this book, however, we don't mention those mysterious physical objects further.

1.3   Forces and interactions

Elementary particles may hit each other. We call these phenomena as **collisions** or **scatterings**. By collisions, they are simply scattered, or merged, or they produce other particles. Those reactions are called together as **interactions** of particles. A particle can sometimes disintegrate to multiple particles. This phenomenon is called **decay**. The decay can be also explained as an interaction of particles; the mother particle, which is the particle before the decay, interacts with the daughter particles, which are the particles after the decay. In the elementary particle physics, all the reactions including decay processes are called **interactions**. The interactions are caused by the **forces** between the particles. Typical interactions are illustrated bellow.

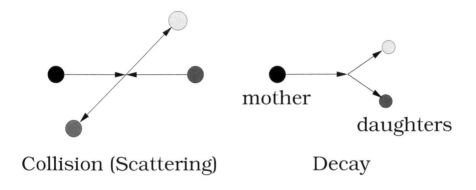

Collision (Scattering)   Decay

1.4 Interactions are carried out by particles

Not only the matters are made of particles, interactions between the particles are also carried out by the particles mediating between them. In the other words, the behavior of the particles are also the outcome of the effect of other particles. So everything in the world is made of a nest of elementary particles.

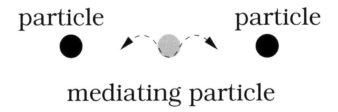

1.5 Types of interactions

Interactions of elementary particles are "basically" classified into three types. I said "basically," because there are extra kinds of interactions concerning Higgs particles, mentioned near the end of this book. The three major types of interactions are **electromagnetic interaction**, **strong interaction** and **weak interaction**.

Electromagnetic interactions are most familiar interactions to us. They appear in our daily life. Everything we observe in the daily life, except the gravitating phenomena, is caused by the electromagnetic interactions. If we hit a wall, we feel ache. This is caused by the electromagnetic interactions. Hydrogen and oxygen may blast to flames. This is also caused by the electromagnetic interactions.

Electricity and magnetism are combined in the modern physics. For example, if there is an electric charge in the space, electric field is generated around the charge. For a person who is moving relative to that charge, the charge appears to be moving. When the charge moves, an electric current flows. This electric current generates magnetic field around it. So the moving person sees the magnetic field as well as the electric field. From this example, we can imagine that the electric field and the magnetic field share the same origin. So we call them together as electromagnetic field.

Interactions between the charged particles are called electromagnetic interactions.

Strong interactions work mainly between quarks. The strong force also confines protons and neutrons in the nucleus. Protons have positive electric charges. Thus they repulse each other by electromagnetic forces. To make them together, the strong forces have to be stronger than the electromagnetic forces.

We do not encounter strong interactions in our daily life. This is due to the reason that the strong forces have very short range. Their range is about the size of a nucleus, far shorter than the scale of the daily life, or even the size of an atom.

Weak interactions are mainly seen in decay processes of nuclei or elementary particles. Disintegration of nucleus occurs if the number of neutrons and the number of protons are unbalanced. Because the neutrons are slightly (1.4%) heavier than the protons, too many neutrons cause too heavy nucleus. Hence it is stabler if one of the neutrons decays to a proton, which makes the nucleus lighter. This decay is carried out by the weak interaction. Opposite to this, if the number of protons far exceeds the number of neutrons in the nucleus, repulsive forces among the protons accumulate. In those cases, it is stabler if one of the protons decays to a neutron. This decay also occurs by the weak interaction.

Time scales of these decays are between several minutes to thousands of years depending on the kind of the decays. Contrary to this, time scale of light emissions from atoms is an order of 1/100,000,000,000 shorter than those decays. The difference between the two reactions is that the nuclear decays are caused by the weak interactions while the light emissions are caused by the electromagnetic interactions. The longer time scale of the nuclear decays is due to the weakness of the weak interaction.

This weakness of the weak interaction causes the prolonging problems of the nuclear accidents at Chernobyl and Fukishima Daiichi. Because, after the accidents, some of the fallout materials are decaying by the weak interactions, and it takes a long time for those materials to decay. Also the reason why the Sun can shine more than 5,000,000,000 years is that the weak interactions are involved in the chain reactions of nuclear fusions in the Sun. The weakness of the weak interactions prolongs the Sun's life enormously.

We rarely encounter the weak interactions in our daily life, other than the above examples. The reason is the short ranginess of the weak interaction. Its typical range is one hundredth of the size of a nucleus.

1.6 Units and numerical expressions

Up to this point, we used very large numbers and very small numbers. We have to use very large or very small numbers to describe the elementary particles. If we used the conventional expressions, we have to write many "zeros." To avoid these cumbersome expressions, we use some special techniques.

One of them is a unit. In our daily life, we mostly use either the International System of units (SI units) or the yard-pound units. Those are designed for the convenience in the everyday life, but they are not particularly convenient for the particle physics. The SI units, for example, are nothing to do with the fundamental physics. One meter, for example, is 1×10^{-4} of the meridian of the earth. As a matter of fact, lengths are not relativistically invariant quantity. In the particle physics, we most often use electron volt units. The electron volt is a unit for energy. Assuming there are a pair of parallel electrodes; one is a cathode and the other is an anode, as shown in the figure below. Then one volt of voltage is applied between the cathode and the anode. We put an electron on the surface of the cathode. By the electric force, the electron starts moving from the cathode with the zero initial speed. Then the electron is accelerated to the anode. One electron volt (eV) is the kinetic energy of the electron when the electron arrives on the anode surface. It should be noted that the final energy is independent of the distance between the cathode and the anode. The electron volt is a particle physics friendly unit because most of the measurements of the particle physics are done using accelerators, which use electric field to accelerate the charged particles.

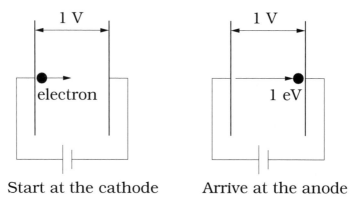

Start at the cathode   Arrive at the anode

The second technique is the usage of index numbers. Even if we use this particle friendly unit, we still need extreme numbers. For those extreme numbers, we use index numbers. For example, 3,000,000,000 is expressed as 3×10^9. The number of zeros is written at the shoulder of 10 in the index number, such as 9 in the above example. For very small numbers such as 1/1,000,000,000, we write a minus number instead. In the case of 1/1,000,000,000, we write it 1×10^{-9}.

The third technique is the usage of large or small units. By using the index numbers, we can write any number without difficulty. However, the index numbers are not easy to pronounce orally. So we use multiples of base unit, such as km for 1000 m. Typical such symbols are listed below. For example, 1×10^{12} eV is expressed as 1 TeV.

Reading	Symbol	value
kilo	k	10^3
mega	M	10^6
giga	G	10^9
tera	T	10^{12}
milli	m	10^{-3}
micro	μ	10^{-6}
nano	n	10^{-9}
pico	p	10^{-12}
femto	f	10^{-15}

Using these symbols, the size of an atom is about 0.1 nm and the size of a nucleus is about 1 fm.

Lastly, we use **natural units** in particle physics. The fundamental constants in physics are the speed of light, c, in vacuum, Planck constant, etc. So in the natural units, we set $c=1$, for example. In this book, however, we explicitly write c since I believe it is more familiar to the readers.

1.7   Structure of this book

In the following three chapters, we briefly describe special relativity, quantum mechanics and field. Those are necessary to explain particle physics. After those, we will present particle classification and three types of fundamental interactions. The goal is to understand the Higgs mechanism.

Chapter 2

Relativity

Basic knowledge of special relativity is necessary to understand the particle physics. In this chapter, we briefly present the special relativity, just enough to read the rest of this book.

Relativity has two kinds: special relativity and general relativity. In this book, if we simply say "relativity," it means the special relativity.

2.1 Mass and energy

Lights travel in vacuum, always at the speed of **light velocity, c**. From this experimental fact, A. Einstein made the **special relativity theory**. A famous formula, resulting from this theory, is $E = mc^2$.

Let me introduce this equation without describing its deriving method. Because there seems to be no simple and easy way to derive this equation.

This equation shows that mass, m, and energy, E, are equivalent with a conversion factor of c^2. As for the conversion factor, light velocity, c, is 3×10^8 m/s. Although the light velocity is a constant, its numerical value depends on the used unit system. The 3×10^8 m/s is a conversion factor in the SI units. Though this number happens to be huge in this particular unit system, we can choose $c = 1.0$ as it is done in the natural units. What I want to emphasize is, that the mass, m, and the energy, E, are equivalent, but the conversion factor is artificial.

$$E = mc^2$$

In our daily life, however, this large conversion factor have a big meaning. It provides the reason why nuclear reactions can produce a large amount of energy. In the nuclear fission, uranium 235 disintegrates into two lighter nuclei and a few neutrons. The sum of the masses after the reaction is significantly smaller than the original mass of the uranium. This difference goes to energy. The energy is huge since c is huge in our typical sense. As a matter of fact, light velocity far exceeds the speed limit of any expressways in the world.

In the world of particle physics, however, conversions between mass and energy are typical phenomena. Light velocity is a typical velocity in the motions of elementary particles.

Let's think about a couple of examples to show the relation between mass and energy. First example is the gravitational potential energy of an object and its mass. If a person lifts a 1 kg mass from the ground to 1 m above, how much is the mass difference in the object before and after the lift. We assume the gravitational acceleration at the earth's surface, $g = 9.8$ m/s^2, is constant and independent of the height.

When the 1 kg mass is lifted upward by 1 m, the gain in its potential energy is

$$U = (1 \text{ kg}) \times (9.8 \text{ m/s}^2) \times (1 \text{ m}) = 9.8 \text{ J (Joule)}.$$

The person, who lifted the mass, looses the same amount of energy, 9.8 J, during this process, due to the energy conservation law. This loss of energy reduces the person's mass while everything else in the earth system, including the 1 kg object, gains mass. The change of mass, ΔM, corresponding to this energy, U, is

$$\Delta M = (9.8\text{J})/(3.0 \times 10^8 \text{m/s})^2 \text{ kg}.$$

By calculating the above, $\Delta M = 1.1 \times 10^{-16}$ kg. The net difference in the earth system's mass, including this person, is zero due to the energy conservation law.

The mass gain in the 1 kg object, Δm, is given by

$$\Delta m = \frac{m}{M_0} \Delta M,$$

where m is the mass of the object ($= 1$ kg) and M_0 is the earth mass: $M_0 = 6.0 \times 10^{24}$ kg. By calculating the above, $\Delta m = 1.8 \times 10^{-41}$ kg. Certainly the gain is extremely small. The present technology is not good enough to measure this difference. But it differs.

It should be noted that g is actually different for the different height. The difference of g by lifting 1 m from the ground can be estimated by 2 × 1 m / (earth radius in m). Since the earth radius is $(1.0 \times 10^7)/(\pi/2)$ m = 6.4×10^6 m, the fractional difference in g is 3.1×10^{-7}. This is far larger than the difference in the mass itself. As a result, by lifting the object by 1 m, the mass of the object increases a tiny amount, but the weight of the object decreases much more.

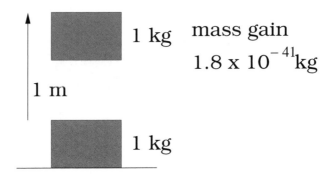

The second example is the relation between temperature and mass. Temperature of an object is basically the kinetic energy of microscopic motions of molecules or atoms inside the object. If we increase the temperature, the object has more internal energy. Hence the mass of the object increases.

Low temperature   High temperature

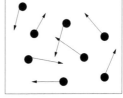

Low mass   High mass

2.2   Energy of a moving particle

We have seen that the energy of a rest particle is the mass of the particle. What is the energy of a moving particle? It is expressed as

$$E^2 = (\boldsymbol{p}c)^2 + (mc^2)^2.$$

Here E is the energy, $\boldsymbol{p}$ is the momentum, m is the mass, c is the light velocity in vacuum. The momentum, $\boldsymbol{p}$, is a vector. If we explicitly write the momentum, $\boldsymbol{p}$, with its (x, y, z) components, $\boldsymbol{p} = (p_x, p_y, p_z)$. Then the above equation is

$$E^2 = (p_x^2 + p_y^2 + p_z^2)c^2 + (mc^2)^2$$

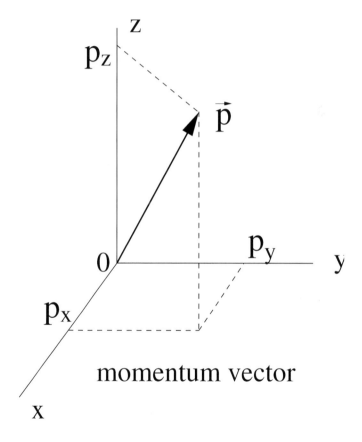

momentum vector

Again, this equation, $E^2 = (\boldsymbol{p}c)^2 + (mc^2)^2$, seems to be not easy to derive. So I just give this formula without explaining how to derive.

If the particle stops, $\boldsymbol{p} = 0$, then $E^2 = (mc^2)^2$. From this result, this formula matches the formula for the rest particles, $E = mc^2$, if we ignore a potential difference in its $\pm$ sign.

Let's summarize this chapter.

- Energy, E, of a particle at rest is equal to its mass, m. The conversion factor between the energy and the mass is c^2, where c is the light velocity in vacuum. If we write this in an equation, $E = mc^2$.

- Energy, E, of a moving particle is expressed by $E^2 = (\boldsymbol{p}c)^2 + (mc^2)^2$ Where $\boldsymbol{p}$ is the momentum of the particle. The momentum of the particle is a vector. If we write the momentum explicitly with its (x, y, z) coordinate components, $(\boldsymbol{p}c)^2 = (p_x^2 + p_y^2 + p_z^2)c^2$.

Chapter 3

Quantum Mechanics

Microscopic objects, smaller than molecules, behave differently from the objects in our daily life. Since they behave differently from the expectations by our common sense, their behaviors seem to be strange to us. The strange behaviors of such objects are described by **quantum mechanics**. In this chapter, we introduce some basic concepts of quantum mechanics that are necessary to read the rest of this book.

3.1 Light and its Energy

Let's think about light in the vacuum. We all know that light is a wave. It diffracts after passing a pair of narrow slits, which is a strong indication of the light as a wave. The light as a wave has wavelength, λ, frequency, ν, amplitude, phase, polarization and the velocity, c.

When we illuminate light on a certain metal surface, electrons are released from the surface. This phenomenon is called **photo electric effect**. In this phenomenon, the light hits an electron in the metal and if the force transformed to the electron by the collision is strong enough, the electron will be released from the metal surface. In this effect, irradiation of light with shorter wavelength can release higher energetic electrons. This indicates that the light with shorter wavelength has higher energy. Brighter light, i.e. the light with larger amplitude, releases more electrons, but their energies are not particularly higher.

The frequency of light, ν, and the wavelength, λ, has a relation

$$\nu = \frac{c}{\lambda}.$$

This relation expresses that the frequency is inversely proportional to the wavelength. This, with the knowledge of the relation between the wavelength and the energy, means that the light with higher frequency corresponds to higher energy. More explicitly, light energy, E, is proportional to its frequency, ν, as

$$E = h\nu$$

where the proportionality constant, h, is called the **Planck constant**.

This can be also expressed using λ as

$$E = \frac{hc}{\lambda}$$

where h and c are constants. The energy, E, is inversely proportional to the wavelength, λ.

3.2   Light is also a particle

If the energy of light is increased, the characteristics of light changes: i.e. visible light → ultra violet → X-ray → gamma-ray. Gamma-rays, emitted from a radioactive isotope, can be counted by an experimental equipment, like the one shown below.

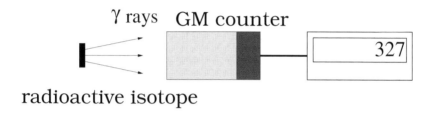

One can count the number of lights. This shows that the light is also a particle. The light as a particle is called **photon**. And a photon is an elementary particle. A light is a wave as well as a particle. This is an example for the **wave-particle duality**. The most basic concept of quantum mechanics is this wave-particle duality. All the elementary particles also behave as waves. Or, in the other words, **microscopic objects are always oscillating**.

wave-particle duality	quantity
wave	amplitude, wavelength, frequency, velocity, polarization, phase
particle	energy, momentum, velocity, spin

3.3   How about electrons and quarks?

Electrons and quarks also have wave-particle duality. This has been shown by double slit experiments, for example. In a double slit experiment, as shown below, an electron, started from the point electron source on the left, travels through the double slit, then it hits the screen located on the right. If the electron is a particle, it has to move on the straight line: one of the two dashed-lines shown in the figure. Then it should hit the point that the line crosses the screen. The experiment shows, however, the electron can hit the other point on the screen. This could happen, if the electron hits the edge of the slit and if it is scattered. However, if we accumulate the data for many electrons, their hit-points show a clear pattern: i.e. a diffraction pattern. The diffraction pattern is caused by the interference of electrons passing through the two slits simultaneously, though each electron is emitted one by one from the source. This indicates that the electron can behave as a wave, and it "interferes by itself."

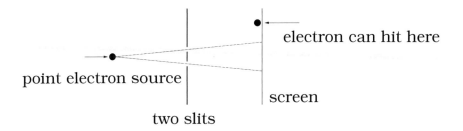

From the diffraction pattern in this experiment, one can derive the relation between the electron's momentum, p, and its wavelength, λ, as

$$\lambda = \frac{h}{p}$$

where h is the Planck constant. The wavelength, λ, is called **de Broglie wavelength**. The momentum of the electron is inversely proportional to the de Broglie wavelength, λ, of the electron.

As we have explained up to this point, the wavelength is inversely proportional to the energy for photons while for electrons the wavelength is inversely proportional to the momentum. It may sounds that the formula are different between photons and electrons. This can be straighten out; since, for photons, energy and momentum are basically the same: $E = pc$, we can say that the wavelength is inversely proportional to the momentum for both the photons and the electrons.

There is, however, a difference between photons and electrons. Since photons have no mass, they can not stop. But electrons can stop since they have masses. When an electron stops, its momentum is zero. So the de Broglie wavelength is infinite. The electron, however, still oscillates. The wavelength of this oscillation is called **Compton wavelength**, λ_c.

$$\lambda_c = \frac{h}{mc}$$

where h is the Planck constant, m is the mass of the particle, c is the light velocity in vacuum. This formula means that the wavelength, λ_c, is inversely proportional to the mass of the particle.

There are two kinds of wavelengths, de Broglie wavelength and Compton wavelength, associated to the same particle. What is the difference between them. The de Broglie wave is really a wave. It propagates and produces diffraction. The Compton wave is an oscillation of the particle itself. It does not propagate or produce diffraction. For slowly moving particles, the de Broglie wavelength is much longer than the Compton wavelength. The difference between those two is illustrated below.

Compton wave length

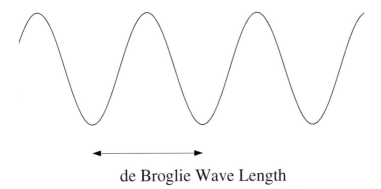

de Broglie Wave Length

3.4   Range determines mass

A particle is a diffused object. It is always oscillating. The oscillation forms a cloud with a diameter determined by the Compton wavelength. The mass of the particle is inversely proportional to the Compton wavelength. From these, we can deduce that the **size of the cloud determines the mass of the particle**.

For example, photons have zero mass. But, if the range of the photon is limited, the photon gets a mass. Assuming the case that we look at the light emitted from an LED lamp on the ceiling, if the distance between the lamp and our eyes is $L = 2$ m, the range of the photon is limited by 2 m, and its mass is calculated using

$$m = \frac{h}{cL}$$

By substituting h=6.6×10^{-34} Js and $c = 3.0×10^8$ m/s, the mass is m = 6.6×10^{-34}/(3.0×10^8×2) = 1.1 ×10^{-42} kg. This is a tiny mass. But it is not zero. We call such a photon with limited range as **virtual photon**. If the range of the photon is a length of our daily life, such as in this case, we can ignore the mass and call the photon as **real photon**. The distinction between the real photon and the virtual photon is not strict. But, in the world of particle physics, a typical range is 1 fm, for which the mass is 200 MeV/c^2. Then, we consider those photons as virtual photons.

As we have seen in this case, if the range of a particle is limited, its range determines the mass of the particle.

3.5 Position spread and momentum spread

Let's go back to the relation between the momentum, p, and the wavelength, λ, of a particle.

$$\lambda \times p = h$$

We can assume that the wavelength is a measure of the particle's cloud. Then, this equation implies that neither the particle's spreading size nor the particle's momentum can get to zero. To make the situation more general, we slightly modified the above equation and write the formula as

$$(\text{position spread}) \times (\text{momentum spread}) = h.$$

We call this formula as the **Heisenberg uncertainty principle**. This principle is one of the most useful principles for understanding the particle physics.

Let's think about the meaning of the Heisenberg uncertainty principle. In the figures, shown below, the left figure illustrates a particle and the right figure illustrates a simple oscillation of a pendulum.

moving around   simple oscillation

Particles are always oscillating even when they are at rest. The oscillation of a particle can be considered as a simple oscillation of a pendulum. So, to understand the particle's behavior, we examine the motion of the pendulum.

When the weight of the pendulum reaches the highest point, A, in the right figure, the pendulum stops. This indicates that when the "position spread" is a maximum, the "momentum spread" is a minimum. When the weight of the pendulum reaches the lowest point, B, in the right figure, the pendulum reaches its top speed. This indicates that when the "momentum spread" is a maximum, the "position spread" is a minimum. From this example, we can estimate that the oscillations have a general characteristics of

$$(\text{position spread}) \times (\text{momentum spread}) = (\text{constant}).$$

This has a strong resemblance to the Heisenberg uncertainty principle. In the Heisenberg uncertainty principle, this constant is h. So, we can say "the Heisenberg uncertainty principle is based on the concept that particles are always oscillating." And the important point is that the oscillations are involved in all the phenomena of particle physics. **Oscillation is the fundamental concept of quantum mechanics.**

3.6 Width and slope of physical variable

The Heisenberg uncertainty principle is also understood as a relation between the width and the slope of the distribution of a physical variable. The figures, shown below, illustrate three distributions with different widths and slopes. Those distributions can be interpreted as the position distributions of a particle.

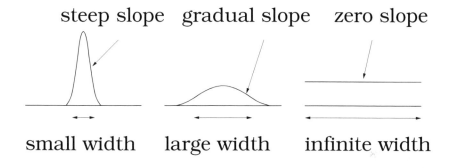

The left figure shows a small width and a steep slope case. The center figure shows a large width and a slow slope case. The right figure shows an extreme case: an infinite width and zero slope. From these distributions, we can figure out a relation: wider the distribution, slower the slope. Slope can be also interpreted as a rapidness of change. So, a steep change corresponds to a narrower location.

$$(\text{width}) \times (\text{slope}) = (\text{constant})$$

To relate the shapes of the distributions to the Heisenberg uncertainty principle, the width corresponds to the "position spread" and the slope, or rapidness of change, corresponds to the "momentum spread." In quantum mechanics, we call such a pair of physical variables as **canonical conjugate**. Position and momentum are canonical conjugate.

3.7   Momentum and energy

In the above arguments, the momentum of a particle was discussed. How about the energy of the particle. Or, what is the difference between energy and momentum. As I know, many high school students have trouble choosing the right variable to describe the magnitude of motion: momentum or kinetic energy.

The biggest difference between momentum and kinetic energy is that the momentum is purely mechanical while the kinetic energy can be converted to anything: mechanical, thermal, chemical, or electrical energies.

To visualize this difference, let's think about the case of a car colliding with a person. If the collision is purely elastic, the person is not damaged at all. He, or she, is not injured or even feels no pain. To make the person injured, or feel pain, a part of the initial kinetic energy has to be converted to distortional, or chemical, or electrical energies. If the amount of the converted energy is large, the damage is also large. This means that, if the person's flying distance is long after the collision, the injury will be small. But, if the distance is short, the injury will be large, because the converted energy is large.

Since kinetic energies can be converted to other types of energies via work, the hit person is injured and he/she feels pain. On the other hand, momentum itself can not be converted to other forms. Hence it does not directly relate to the damage.

The momentum has its direction. A variable with direction is called vector. So, the momentum is a vector. In the Heisenberg uncertainty principle, the canonical conjugate of a vector has to be also a vector. For the momentum case, the canonical conjugate is a position which is also a vector. If we write a vector using its components, the momentum, p, is written as (p_x, p_y, p_z), where p_x is the x-component of the momentum. A position vector, x, is written as (x, y, z). Then the Heisenberg uncertainty principle is written as

$$\begin{aligned}
(x)(p_x) &= (h\text{:constant}) \\
(y)(p_y) &= (h\text{:constant}) \\
(z)(p_z) &= (h\text{:constant}).
\end{aligned}$$

The canonical conjugate of energy is time.

The following is a summary of this chapter.

- The energy, E, and the wavelength, λ, of a photon satisfy the relation: $E = hc/\lambda$, where h is the Planck constant, c is the velocity of light in vacuum.

- Elementary particles have wave-particle duality.

- The basic concept of quantum mechanics is oscillation. In the microscopic scale, everything oscillates.

- Elementary particles are diffused objects. The range of their spreading is Compton wavelength, λ_c, which satisfies the relation: $\lambda_c = h/(mc)$.

- The magnitude of the momentum, p, and the de Broglie wavelength, λ, of a particle satisfy the relation: $\lambda = h/p$.

- The position spread and the momentum spread of the elementary particle are described by the Heisenberg uncertainty principle as (position spread) × (momentum spread) = h.

Chapter 4

What is Field?

In this chapter, the concept of "field" is introduced, which is one of the three basics of the particle physics: i.e. the relativity, the quantum mechanics, and the field.

4.1 Electric field and magnetic field

Let's think about space. First, the space provides a place that a particle can stay. Also the space is "filled with vacuum." But the vacuum itself seems to be nothing. The particle physics, however, tells us that vacuum has physical quantities. Vacuum has varieties of characteristics. After all, the space is not nothing at all.

To see this space's characteristics, if we place an electric charge in the space. The charge generates electric field around it. This electric field is a feature of vacuum. There are differences between the vacuum with or without electric field. This is also true for the magnetic field.

Since vacuum can have the electric and magnetic field, we may be convinced that vacuum is something after all.

4.2   Field also oscillates

As described in the quantum mechanics chapter, all the physical objects oscillate in the microscopic scale. Field is also a physical object. It is not an exception. Field also oscillates.

Since the entire Universe is filled with vacuum that can have field, there are oscillators everywhere in the Universe. The amount of energy from these oscillations is estimated to be vast. This tremendous amount of energy in vacuum could produce a significant effect to the Universe. This effect could be many orders of magnitude larger than the effect of the "dark energy." Why this vacuum energy of the field oscillations does not have significant effect to the Universe? We do not know the answer yet. The answer is left for you.

The oscillation of field can be considered as a simple harmonic oscillation. Actually, any oscillation can be decomposed to a mixture of many simple harmonic oscillations. An example of the simple harmonic oscillation is a spring motion, consisting of a weight of mass, m, attached to a spring with a spring constant, k.

The kinetic energy, E_k, of the spring oscillation is given by

$$E_k = \frac{1}{2}mv^2.$$

The potential energy, stored in the spring, is

$$V = \frac{1}{2}kx^2.$$

Total energy, E, is the sum of the kinetic energy, E_k, and the potential energy, V, namely

$$E = \frac{1}{2}mv^2 + \frac{1}{2}kx^2.$$

The period of oscillation, T, is given by

$$T = 2\pi\sqrt{\frac{m}{k}}.$$

By taking the squares of the both sides in the above equation,

$$T^2 = (2\pi)^2 \frac{m}{k}.$$

Since the frequency, ν, is the inverse of T,

$$\nu^2 = \frac{1}{(2\pi)^2}\frac{k}{m}.$$

Using this, the spring constant, k, is expressed by the frequency, ν, as

$$k = m(2\pi\nu)^2.$$

By substituting k into the total energy equation,

$$E = \frac{1}{2}mv^2 + \frac{1}{2}m(2\pi\nu)^2 x^2.$$

Since the momentum, p, is written as,

$$p = mv$$

the total energy can be written using, p, instead of v, as

$$E = \frac{1}{2}\frac{p^2}{m} + \frac{1}{2}m(2\pi\nu)^2 x^2.$$

By dividing both sides with E, and swapping the right side and the left side,

$$\frac{p^2 + (2\pi\nu m)^2 x^2}{2mE} = 1.$$

This is an equation of ellipse with x for the x-axis, and p for the y-axis. The equation of ellipse in the standard form is

$$\frac{x^2}{a^2} + \frac{y^2}{b^2} = 1,$$

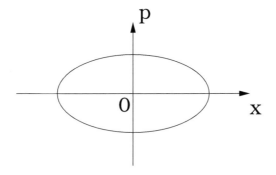

where a is half the diameter along the x coordinate, b is half the diameter along the y coordinate. In the equation of the spring oscillation, a

corresponds to the maximum deviation in position and b corresponds to the maximum deviation in momentum.

Then, we return to the Heisenberg uncertainty principle as

$$\text{(deviation of position)} \times \text{(deviation of momentum)} = h.$$

If we apply this to the spring motion,

$$
\begin{aligned}
a &= \text{(deviation of position)} \\
b &= \text{(deviation of momentum)},
\end{aligned}
$$

then

$$ab \sim h$$

This relation is basically equal to the area of the ellipse, S, which is given by

$$S = \pi ab$$

thus the Planck constant corresponds to the area of the ellipse.

For the spring motion,

$$
\begin{aligned}
a &= \sqrt{2mE} \\
b &= \frac{\sqrt{2mE}}{2\pi\nu m}.
\end{aligned}
$$

By substituting these in the above equation, the area, S, of the ellipse for the spring motion is

$$S = \frac{E}{\nu}.$$

By setting this area as the Planck constant,

$$S = h,$$

we get

$$E = h\nu$$

which is exactly the same as the formula for the photon's energy and its frequency. From this, we can deduce that the photon is an oscillation in vacuum. Or, an oscillation in vacuum is identified as a photon.

To summarize the arguments up to this point, first, vacuum has field. The field is a physical object. Since all physical objects oscillate, the field also oscillates. And this oscillation can be identified as a photon.

Up to this point, we only introduced electric and magnetic field. But other elementary particles also have field. Electrons have "electron-field," quarks have "quark-field." Oscillations of the electron-field produces electrons. Oscillations of the quark-field produces quarks. **Every elementary particle is an oscillation of its field.** When the oscillation ceases to its minimum, the particle disappears and the field returns to the calm vacuum.

The figure, shown below, illustrates the mechanism that the electron-field near the center is excited to an electron. The each point in the vacuum has an electromagnetic field, an electron-field, a positron-field, a quark-field, etc.. When one of the fields is excited, that field starts oscillating and the particle, corresponding to that field, appears.

electron

electro-magnetic field		electro-magnetic field		electro-magnetic field			vacuum
electron field	positron field	electron field	positron field	electron field	positron field		

As a summary in this chapter,

- Vacuum is not an empty space.

- Vacuum has field which is a physical object.

- Since all physical objects oscillate, field can also oscillate.

- Each elementary particle has its corresponding field, or vice verse.

- An oscillation of the field creates an elementary particle corresponding to the field.

- When the oscillation gets its minimum, the particle disappears, remaining only the calm vacuum.

Chapter 5

Antiparticles

Elementary particles are classified into two types: particles and antiparticles. In this chapter, antiparticles will be explained based on the **Dirac's hole theory**.

5.1 Dirac equation and positron

From the relativity, energy, E, momentum, $\boldsymbol{p}$, mass, m, satisfy,

$$E^2 = (\boldsymbol{p}c)^2 + (mc^2)^2.$$

Since this equation is a quadratic sum form, we may take a square root of this equation. Then the results are

$$
\begin{aligned}
E &= +\sqrt{(\boldsymbol{p}c)^2 + (mc^2)^2} \\
E &= -\sqrt{(\boldsymbol{p}c)^2 + (mc^2)^2}.
\end{aligned}
$$

There are two solutions: one has a minus sign and the other has a plus sign. Both the equations satisfy the original quadratic equation.

In the above equations, when the particle stops, i.e. $p = 0$, then

$$E = +mc^2$$
$$E = -mc^2.$$

Since the mass is positive $m > 0$, the second equation in the above has a negative energy, $E < 0$. Here we may pick up only the $E > 0$ case, because the $E < 0$ case has no physical meaning. However, we can not do this. To explain the reason, why we can not discard the $E < 0$ case, we need a little more explanation.

P. A. M. Dirac tried to discompose the quadratic equation, shown below, into two linear equations.

$$E^2 - (pc)^2 - (mc^2)^2 = 0$$

The reason, why Dirac wanted a linear equation, is that he thought fundamental equations should be linear in energy.

Before we try to discompose the above equation, we try a simpler case as an exercise

$$x^2 + y^2 = 0.$$

This equation can be factorized using the imaginary unit, $i = \sqrt{-1}$, as

$$(x + iy)(x - iy) = 0.$$

The equation of the energy, momentum, mass, is more complex than this. First, the momentum is a vector. If we write the equation with writing each component of the momentum vector explicitly, it is

$$E^2 - (p_x c)^2 - (p_y c)^2 - (p_z c)^2 - (mc^2)^2 = 0.$$

Dirac assumed that the above equation can be factorized in the form

$$(E - \alpha_1 p_x c - \alpha_2 p_y c - \alpha_3 p_z c - \beta mc^2)(E + \alpha_1 p_x c + \alpha_2 p_y c + \alpha_3 p_z c + \beta mc^2) = 0$$

where $\alpha_1, \alpha_2, \alpha_3, \beta$ are constants.

Are there any such constants really exist? Dirac found such constants. They turned out to be 4×4 matrices. So, using 4×4 matrix constants, $\alpha_1, \alpha_2, \alpha_3, \beta$, he could factorize the quadratic equation.

We call one of the above two linear forms as **Dirac equation**. It can be either

$$E - \alpha_1 p_x c - \alpha_2 p_y c - \alpha_3 p_z c - \beta mc^2 = 0,$$

or

$$E + \alpha_1 p_x c + \alpha_2 p_y c + \alpha_3 p_z c + \beta mc^2 = 0.$$

After finding the 4×4 matrices, Dirac studied the meaning of the 4×4 matrices. He found that 4 is decomposed into

$$4 = ((E > 0) + (E < 0) = 2) \times (("spin"-up) + ("spin"-down) = 2).$$

As for the "spin", we'll explain it later. Here, we just use that the spin requires the factor 2. Since the constants have to be 4×4 matrices, the energy has to have two states: $E > 0$ and $E < 0$. So, we can not ignore the $E < 0$ case. The important point is, if we ignore the $E < 0$ case, 4×4 matrices, αs and β, can not be used, hence there's no Dirac equation.

Since the Dirac equation requires $E < 0$, we could conclude that such equation is invalid, i.e. no valid equation should require $E < 0$. An another possible way to avoid this difficulty is that the energy gap between $E > 0$ and $E < 0$ may work as a barrier for the electrons to fall into $E < 0$. Positive energy electrons can lose energy, by slowing down, to the minimum: $E = mc^2$, but they can not go down below because there's an energy gap between $E = +mc^2$ and $E = -mc^2$. However, this explanation does not work. Since electrons are always oscillating, they can oscillate through the energy gap and go to the negative energy state. It is like the waves of ocean water, that can sometime go over the bank, though it is rare to occur. For the electron case, all electrons eventually fall into the negative energy state. And, in the negative energy state, when electrons lose their energy, they move faster. So, they fall into infinitely "large" negative energy.

Dirac solved this difficulty, or we might say that he came up with a crazy idea. His idea is called **hole theory**. Since electrons are true matters, only one electron can be placed at one point. Imagine a soccer ball is placed at the corner of the soccer court. No other soccer ball can be put at the same place. In the case of electrons, electrons are always oscillating. Hence, no other electron can be placed inside the oscillating volume of the electron. This phenomenon is called as **Pauli's exclusion principle**.

Dirac used this Pauli's exclusion principle to solve the negative energy electron problem. He proposed that if everywhere is filled with negative energy electrons: "negative energy sea," no more electron can fall into the negative energy sea. If an ordinary person says such a crazy idea, people think that he is just joking. Nobody takes it seriously. Dirac, however, believed his theory. And the nature followed him. If we assume the Dirac's theory, i.e. the hole theory, is correct, it also predicts the following. If a high energy gamma-ray hits an electron in the negative energy sea, giving enough energy to the electron, the electron can jump to a positive energy state. After the negative energy electron is ejected from the negative energy sea, there will be a "hole," left in the sea. This hole behaves exactly the same as the electron except the sign of its electric charge. The hole has a positive charge, opposite to the ordinary electron. This phenomenon can be seen to us as a pair production of an electron and a positive charge electron. Shortly after the Dirac's proposal, such a positive charge electron was discovered in the cosmic-rays.

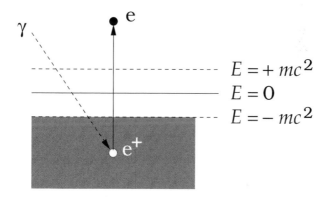

The positive charge electron is now called as **positron**. A particle is called antiparticle, if it has a characteristics, exactly the same as an ordinary particle except the sign of its electric charge. The positron is an example of the antiparticle.

The concept of negative energy sea is almost the same as the concept of the "field." The "sea" and the "field" are basically the same. They fill everywhere in the space, or vacuum. So, Dirac was the first person who almost discovered the concept of particle field, the electron field in his case.

5.2 Antiparticles

Basically every particle has its antiparticle. The comparisons of particle and antiparticle are summarized in the following table.

physical quantity	particle	antiparticle
electric charge	Q	$-Q$
mass	m	m
energy	E	E

For electrically neutral particles, existence of antiparticle is not trivial. For photons and Z^0s, antiparticles are themselves. The Z^0 will be explained near the end of this book. For neutrinos, the anti-neutrinos could possibly be the neutrinos themselves.

5.3 Other explanation for antiparticles

The hole theory is a revolutionary idea and it is also useful to explain semiconductors as well as antiparticles. The hole theory, however, is not the only way to explain antiparticles.

J. Wheeler and R. Feynman proposed the idea that antiparticles are the particles moving backward in time. Based on this idea, Feynman invented a graphical way of describing the elementary particle interaction, **Feynman diagram**. In Feynman diagrams, particles and antiparticles are indicated by the arrows, pointing to the future and the past, respectively. An example of the Feynman diagram is shown below.

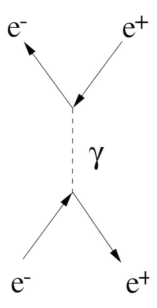

The above diagram shows the reaction that an electron and a positron collide, and they change to a photon, then the photon changes to an electron and a positron again. In this diagram, time advances from the bottom to the top. Since electrons are particles, they propagate from the past to the future. Hence, we draw the arrows pointing to the future. Positrons, on the other hand, are antiparticles, they propagate from the future to the past. Hence, we draw the arrows pointing to the past. Typically there are two ways to draw Feynman diagrams; the direction of time is indicated from the bottom to the top, or from the left to the right. In this book, we use the first scheme.

The photon in the above diagram is shown by a dashed line. Photons are usually shown by wavy lines. Here we use the dash line, simply because it is easier to draw.

The vertices in the diagram indicate the interactions. In this diagram, an electron, a positron and a photon interact at each vertex. Since the above interaction is caused by the electrical charge of the electrons and positrons, the type of the interaction in this diagram is the electromagnetic interaction.

Another way of explaining antiparticles is the one using the field. In this view, an electron is an excitation of the electron field, and a positron is an excitation of the positron field. The electron field oscillates with positive frequencies, and the positron field oscillates with negative frequencies. This method does not have to use the Pauli's exclusion principle, as it was used in the hole theory.

The above three methods of describing antiparticles are complementary. They can be used interchangeably.

As a summary,

- To factorize the relativistic equation of energy, E, momentum $\boldsymbol{p}$, and mass, m,

$$E^2 = (\boldsymbol{p}c)^2 + (mc^2)^2$$

  there appear to be two cases: $E > 0$, and $E < 0$. Both those two cases are necessary to factorize the equation. The negative energy case indicates the existence of antiparticles.

- In the hole theory, all the negative energy sea is assumed to be filled with negative energy electrons. A hole in the sea can be considered as a positron — an antiparticle of the electron.

- Antiparticles can be considered as the particles moving backward in time.

- Antiparticles can be also considered as excitations of the antiparticle field, so as particles are excitations of the particle field.

Chapter 6

Particle Classification

Elementary particles are defined as the particles which can not be divided into smaller parts, at least in the current knowledge. Electrons and quarks are elementary particles. Protons and neutrons are not elementary particles any more since we know that they are made of quarks. In this chapter, however, we call protons and neutrons as elementary particles, which is convenient to explain the historical development of the particle classification.

6.1 History of elementary particle's discovery

The first elementary particle, discovered by J. J. Thomson near the end of 19th century, was the electron. Then, the proton was discovered by E. Rutherford, and the neutron was discovered by J. Chadwick. Those three, and the photon, were the only elementary particles discovered before the end of the World War II.

After the World War II, nuclear physics received big financial supports as well as big attentions, because of the successful development of the

nuclear weapons during the war. In the United States, large cyclotrons
and synchrotrons were built, which enabled nuclear physicists to create
new elementary particles. In Europe and in Japan, cosmic-rays were
used to study new elementary particles because of their poor funding.

As a result of these active studies, many elementary particles were
discovered. Among the particles, discovered in those early years, are
muons, π-mesons (pions), K-mesons (kaons), Λ, Δ. During those early
developments, the classification of particles was a big issue, similar to
the case in the biology in the 19th century.

The first such attempt was done by W. Heisenberg before the World
War II. He proposed that proton and neutron are just a different aspect
of the same particle, **nucleon**. Before getting into the details, let's look
at the properties of the proton and the neutron.

name	symbol	mass	electrical charge	spin
proton	p	938.3 MeV/c^2	$+e$	1/2
neutron	n	939.6 MeV/c^2	0	1/2

The masses of the proton and the neutron differ only 0.14%, as seen
in the above table. Though it is slightly strange that the neutron with
no electric charge is heavier than the proton with positive charge,
those two could be considered as the variations of the same particle,
i.e. nucleon. He proposed a new variable, **isospin** (originally called
"isotopic spin"). According to him, the nucleon has the isospin, and
if the isospin is pointing upward, the nucleon becomes a proton, and
if the isospin is pointing downward, the nucleon becomes a neutron.
What is the "spin" and the "isospin?"

6.2   Spin and isospin

Electrons have electric charge. In addition, electrons are tiny magnets
as well. The electron has an N pole and an S pole, similar to the case of

the Earth. Since there are two poles, N pole and S pole, we call such an object as a **dipole**. The strength of the electron's dipole is explained by a model that the electron has a loop of electric current. Assuming that the current is generated by the electron's charge moving with the light velocity, c, on the circle with the diameter of its Compton wavelength, this current gives the strength of the magnetic field of the dipole. From this, we think that electrons are rotating. This rotation is called **spin**. Since electrons are always rotating and they never stop, we say that electrons have spin.

We define the direction of a rotation using the axis of the rotation. Then, to define its rotating direction, we use the moving direction of the right handed thread when turned in a clockwise direction.

The rotation direction, or spin direction, of an electron has a peculiarity. If we put an electron in the magnetic field, the spin direction can have only two selections: either parallel or anti-parallel to the magnetic field. The spin can not aim to $45°$ to the direction of the magnetic field, for example. It can only aim to either $0°$ or $180°$ to the field. Since it has only two directions, we call the spin directions as "up-spin" for the parallel, and "down-spin" for the anti-parallel.

Electrons are not the only elementary particles that have spin. Quarks and some other elementary particles also have spin. We come to this point later.

The electron's spin has an another strange behavior. The electron has to turn twice, $720°$, to return to its original state. For a rotation in our daily life, we only have to turn $360°$ to return to the original state.

Another special character of the electron's spin is that it is a three-dimensional (3D) rotation. What is a 3D rotation? Or, what is the dimension? One dimension (1D) means a line. In 1D, an object can move only on the line. A motion in 1D can be expressed by only one parameter, namely x-coordinate. There's only one freedom, back and forth, in 1D. Rotations are not possible in 1D. In 2D, an object can move two independent directions, x and y coordinates. It can move on a surface. Rotations are possible in 2D. In the 2D rotations, we can not set the directions of their axis. One can only choose the center of the rotation and the rotating direction: either clockwise or counter-clockwise. In 3D, an object can move three independent directions, x, y and z coordinates. With increasing the dimension, more complicated motions can be possible. In 3D, the number of freedoms in motion is three. Rotations are, of cause, possible. In the 3D rotations, we not only can select the centers of the rotations, but also we can select the directions of their axis. The figures, shown below, illustrate typical 2D and 3D rotations.

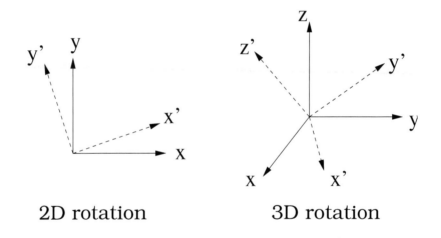

2D rotation   3D rotation

Next, we change the subject to a complex number. As you had learned in the high-school, complex numbers are useful to solve the quadratic equations. When we have to calculate a square-root of a negative number, we can write the solution using an imaginary unit, $i = \sqrt{-1}$. The i is also very useful in the elementary particle physics.

In complex numbers, there are two independent parts: a real part and an imaginary part. So, the complex number can represent 2D. A complex number can be shown by a point on a plane — the complex plane. One can also rotate complex numbers, as shown below.

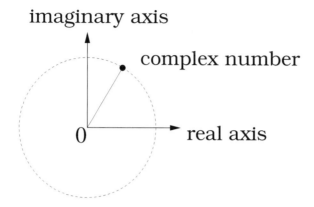

The electron's spin is expressed using complex numbers. The electron's spin can have only two directions: up or down. So, you may think that the spin is a 2D rotation. However, since it is expressed using complex numbers, it has an extra freedom of motion. Thus the number of freedoms becomes three. The 3D nature of the electron's spin will be explained later.

As for the reason, why we have to rotate the electron by $720°$ to return to its original state, it is a consequence of the complex number. For a complex number, it has to be operated by its conjugate to get the real number. The conjugate of a complex number is derived by flipping the sign of the imaginary part of the complex number. Because of this extra operation, the rotation angle will be doubled. This causes the electron to rotate twice to get the original state. You may simply understand this, that to get a real number, 1, a multiplication of the imaginary unit, $i = \sqrt{-1}$, is necessary as $1 = (-i)(i)$. So, two turns are necessary to get the original state.

6.3 Origin of electron spin

Let me change the subject in this section. We'll think about how the electron has its spin. Electrons are considered as point particles. All the evidences coming from the scattering experiments and the other measurements are consistent with the electrons as point particles. Electrons are magnets as well as charged particles. Since we can not extract magnetic charges from the electrons, we consider the electrons are electromagnets. In order to be an electromagnet, electric current has to flow on a loop. This means that the electron has to have finite size, not a point. The electromagnetic theory tells us that the strength of an electromagnet is determined by the product of the strength of the current and the area of the current loop. As a matter of fact, the strength of the electromagnet of electron is consistent with a model that the current is flowing on the circumference of a circle with its radius of the electron's Compton wavelength and with the current strength as the electron's charge flowing with the light velocity.

Why the sizeless electron can have such a current loop? This is similar to the situation as the Compton wavelength. Though the electron is a point particle, it has a finite spacial spread given by the Compton wavelength. This is due to the quantum oscillation, or in the other words: the Heisenberg uncertainty principle.

Each electron has its own specific direction. The reason for this is not known yet. It could be coming from the structure of the electron at a very small scale. An electron may be a tiny string or a tiny bar. If the electron is a tiny string, for example, it can have a direction. Though we don't know the reason yet, electrons have their specific directions.

Once the electron has its specific direction or axis, it obtains a freedom to rotate around the axis. If there is a freedom for a motion, an oscillation always takes place for that freedom. This is due to the quantum nature of tiny systems. Because of this, rotating oscillations are accompanying with the electrons as quantum oscillations.

Quantum oscillations are aroused for all the pairs of canonical conjugate quantities. In this case, the pair is angular momentum and its rotation angle. The Heisenberg uncertainty principle for this pair is written as,

$$(\text{angular momentum} : L) \times (\text{rotation angle} : \theta) = \hbar.$$

For ordinary rotations, rotating motions become identical if the angle is changed by 360°, or one turn. This makes a restriction on the rotating angle. If a restriction is set to a quantity of a quantum pair, the other quantity of the pair, i.e. a canonical conjugate of the quantity, will resonate. It will be a standing wave. Because of this, the magnitude of the canonical conjugate has to take discrete values. In the rotation case, the angular momentum is the canonical conjugate of the rotation angle. Since the rotation angle has the 360° rotation restriction, the angular momentum has to have a multiple of $\hbar$. The lowest value of this is simply $\hbar$, which we call spin 1.

In the electron's case, the period of rotation is two turns, or 720°, as shown in the figure below. Because of this, the canonical conjugate, i.e. the spin, will be a half of the ordinary rotation. This gives spin 1/2 to the electron.

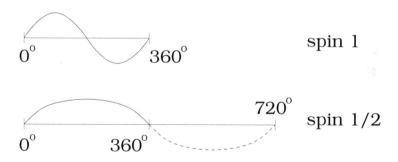

To summarize the story, the origin of electron's spin is (1) electron has its specific direction, or axis, (2) electron returns its original state by rotating two turns, or 720°.

Because of this, as combined with its accompanying quantum oscillation, the election has spin 1/2.

6.4 Unitary transformation

In quantum mechanics, there's a concept called as **unitary transformation**. We briefly explain this concept.

First, what is transformation? The transformation is equal to "change." For example, "moving 3 cm to the right" is a transformation. A rotation by 30°, in the clockwise direction around a certain point, is an another example of the transformation.

As for the word, "unitary," "unit" means a constant with a magnitude of 1. From this meaning, one can imagine that it is a transformation without changing its magnitude. A transformation with a fixed magnitude is a rotation. So, the **unitary transformation is a generalized rotation**. The meaning of the "generalized" is that it can be a phase rotation, or a rotation of states, or the standard rotation.

A phase rotation is a rotation in the complex plane, as shown in the figure (a) below.

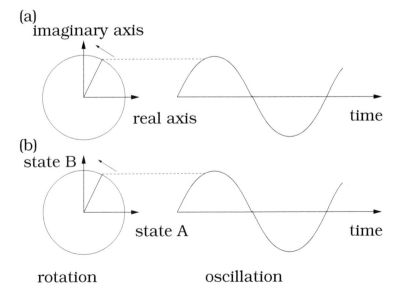

The rotation of states is a phenomena that if there are two states, state A and state B, for example, the two states can be mixed by rotating the axis of the two states, as shown in the figure (b) above. The state A can be a spin up state and the state B can be a spin down state, for example. Then, by rotating 45° from the state A, for example, state A and B will mix with an equal weight.

Rotations are periodic motions, thus, they can be interpreted as oscillations, or waves, as shown in the above figures. A rotation with a constant speed is considered as a harmonic oscillation.

We said the electron's spin is a 3D rotation. What does it mean? If the electron's spin is a 2D rotation, it can be only up or down. The electron's spin, however, can be a mixture of up and down states. This is the indication of the 3D nature of the electron's spin. If the rotation angle inclines, up-spin and down-spin will mix, instead of rotating around the new axis. So, the mixing is the consequence of the rotation. The spin is an example of the unitary transformation. The reader may have hard time to understand the electron's spin. Actually, no physicist can truly understand the electron's spin. They are satisfied only by understanding the mathematical behavior of the spin.

We'll come back to the 2D and 3D rotations later.

Now we change the subject to the **isospin**. Heisenberg noticed a similarity between the nucleon's mass split and the splitting of atomic spectra due to the electron's spin direction. Hydrogen atoms, for example, will emit light when electrical current is discharged in the hydrogen gas. The emitted light is made of a series of specific wavelengths which we call a **line spectrum**. When the gas is in the magnetic field, the line spectrum shifts differently for the spin-up and spin-down cases. This indicates that the energy of the hydrogen atom is slightly different between the spin-up and spin-down cases. Since energy is basically equal to mass, the mass difference between the neutron and the proton can be considered as the energy splitting due to the direction of spin, or "isospin" in this case. So, Heisenberg thought that the difference in the masses of neutron and proton is caused by the splitting due to the difference of the isospin directions. Since this is a generalized rotation, the isospin rotation is also an example of the unitary transformation.

Though Heisenberg used the isospin to explain the difference in the masses between the neutron and the proton, the actual meaning of the isospin is the equality of the two masses. If there's no difference in mass between the neutron and the proton, we can say that the strong interaction is same for the neutron and the proton. For this phenomenon, we say that the **strong interaction has the isospin symmetry**. The symmetry means that there's no difference for changing something, the isospin in this case. The German flag, for example, has a right-left symmetry since it is exactly the same after flipping the right-side and the left-side of the flag.

The isospin symmetry is more than just a 2D rotation or a simple flipping of two states. It is a symmetry for 3D rotations. It is astonishing for neutrons and protons that they have such a 3D rotation symmetry. The reason of this isospin symmetry is base on the nature of quantum mechanics. The quantum mechanical systems have unitary symmetries as their bases due to their oscillations — generalized rotations. Because of this, if there's no difference in the behavior of strong interactions as well as in the masses between the neutron and the proton, they should have a unitary symmetry. Since the neutron and the proton are two states, they are 2D. In addition to this 2D, they are also complex numbers. This adds an extra one freedom to the 2D rotation. As a result, we get the isospin symmetry as a 3D rotation, similar to the case of the spin.

In reality, the neutron is slightly heavier than the proton. Their masses are different. So, the isospin symmetry is not an exact symmetry. It is an approximate symmetry. If a symmetry is not exact, we say "symmetry is broken." Hence, isospin symmetry is slightly broken due to the difference in the masses between the neutron and the proton.

6.5 Group

We will explain a mathematical concept of **group**. We start with back-and-forth motions in 1D. As an example of a back-and-forth motion, we consider the following motion.

(move 3 m forward) + (move 2 m backward) = (move 1 m forward)

This equation describes that moving 3 m forward followed by moving 2 m backward is equal to moving 1 m forward. The action, like "moving 3 m forward," is called an **operation**. Then, the above equation can be written as,

(operation 1) + (operation 2) = (operation 3).

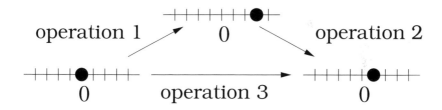

This shows that two operations can be combined into one operation. A set of operations, like these, forms a "group."

A set of 2D rotations forms a group too. For example,

(rotate clockwise by 30°) + (rotate counter clockwise by 40°) = (rotate counter clockwise by $10°$).

A set of 3D rotations also forms a group.

As you can imagine from its name, group is useful to classify the elementary particles. It also plays a big role to understand the dynamics of both the strong and weak interactions.

6.6 SU(2) symmetry

Spin rotations (also isospin rotations) form a group called **SU(2)**. The name, SU(2), comes from "S" for **S**pecial, "U" for **U**nitary and "2" for **2** dimension. The special means that the transformations exclude common phase rotations or flipping. The unitary means that the transformations are generalized rotations, including the rotations in the complex number plane. The 2 dimension, in this case, means that two states are involved: spin-up and spin-down. The point is that SU(2) is a symmetry for 3D rotations of two states. The transformations of SU(2) form a group. This group behaves similarly to the rotations in the 3D space.

6.7 Pions

Pions are introduced by H. Yukawa as the particles that attract protons and neutrons in the nucleus. Pions with electrical charge, which are called charged pions $(\pi^\pm)$, are later found by C. F. Powell and his colleagues in the cosmic-rays. Pions also have neutral types, named as neutral pions (π^0). Neutral pions were found by W. Panofsky. We apply the isospin to classify these pions. To do this, we consider the reaction that two nucleons interact each other by exchanging a pion, as shown in the figure below. In this figure, the nucleons are shown by arrows and the pion is shown by a dash line. The isospin-current in this reaction flows from one nucleon to the other through the pion.

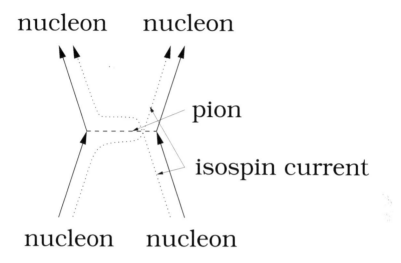

On the pion line, two isospin currents flow in the opposite directions. Two states, i.e. up and down, of one isospin are mixed with the other two states. This mechanism is expressed by

$$(2 \text{ states}) \times (2 \text{ states}) = (3 \text{ states}) + (1 \text{ state}).$$

2 x 2 = 3 + 1

This equation is read as; the two-state of an isospin is multiplied by the two-state of the other isospin, resulting a three-state and a single-state. Since pions have three types: π^+, π^- and π^0, the isospin three-state in the above is suitable to describe pions. As for the single-state, a meson called η can fit this state.

So, we could classify nucleons and pions using the isospin. Those were the particles known in the early 1950s in addition to the particles with no strong interaction: electrons, photons, muons and neutrinos. After this success in the classification, E. Fermi and C. N. Yang proposed a model that the pion is a bound state of a nucleon and an anti-nucleon. For example, π^+ is made of a proton and an anti-neutron. Using this model, we can describe the interaction of a proton and a neutron by exchanging a pion, as shown below.

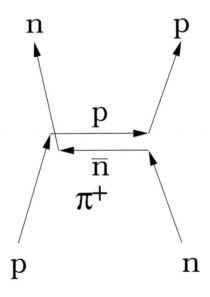

In the above figure, the pion is illustrated as a particle made of a proton and an anti-neutron. The bar above the letter, n, indicates the antiparticle, the anti-neutron in this case.

This model looks attractive since we can save the number of true elementary particles. However, there is an unnatural point. Since the mass of the proton is 938.3 MeV/c^2 and the mass of the neutron is 939.6 MeV/c^2, the combined mass is 1877.9 MeV/c^2. But the mass of the pion is only 139.6 MeV/c^2. If the pion is made of a proton and an anti-neutron, it is natural for the pion to have its mass near the combined mass of 1877.9 MeV/c^2. To make the pion mass of 139.6

MeV/c^2, there has to have a heat loss of $Q = 1877.9 - 139.6 = 1738.3$ MeV, or 93% of the original mass is lost by heat, at combining the proton and the anti-neutron. Only 7% remains as the pion mass. This is possible, but it is highly unnatural.

6.8 Strangeness

During the 1950s, new types of elementary particles, called "V-particles," were found in the cosmic-rays. The figure, shown below, illustrates a pair of V-particles, found in a cloud chamber.

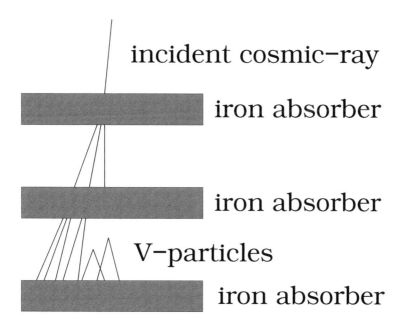

To search for new particles, the cloud chamber was placed at the high mountain to get a large flux of high energy cosmic-rays. The several layers of iron absorbers were inserted in the cloud chamber to make the incident cosmic-rays to interact in them. In the above event, the incident cosmic-ray interacted in the top iron absorber and produced a several secondary particles. One of the secondary particles interacted in the middle iron absorber and produced a pair of V-particles. Those were called V-particles because the shape of the observed tracks looks like an upside-down V. What they called V-particles themselves are not visible because the V-particles are electrically neutral. Only their decay particles are observed as upside-down V's.

Though this kind of interaction was quite rare, when it happened, V-particles were produced frequently as a pair, as illustrated in the above figure. There was no known particle like these at the time of this discovery. They were indications of new particles. In addition, it was strange that why those particles were produced by pairs. Because of these, V-particles were also called **strange particles**.

M. Gell-Mann and K. Nishijima independently noticed that, if the strange particle has a quantity which conserves at their production, the strange particles must be produced by pairs. Gell-Mann called this quantity as **strangeness (S)**. He assumed that one of the V-particles has $S = +1$ and the other V-particle has $S = -1$ and the ordinary particles (non-strange particles) have $S = 0$. Then to conserve strangeness at the production,

incident ordinary particle $(S{:}0) = $ V-particle $(S{:}{+}1) + $ V-particle $(S{:}{-}1)$.

Thus the V-particles have to be produced by pairs.

To explain the V-shapes, Gell-Mann and Nishijima also assumed that the strangeness conserves only in the strong interactions; weak interactions do not have to conserve strangeness. Then the long life of the V-particles can be explained if the V-particles can not conserve strangeness at decay, thus they have to decay weakly. This can happen if there is no other strange meson, lighter than the V-particles, for example. The following figure illustrates the production and the decays of V-particles.

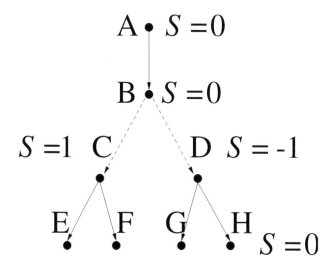

In the above figure, the incident cosmic-ray, A, with negative electric charge, hits a proton, B, in the iron nucleus. Here A and B are ordinary particles, thus $S = 0$ for both A and B. By this collision, two V-particles, C and D, are produced. C is a neutral particle with $S = +1$ and D is also a neutral particle with $S = -1$. Hence A + B → C + D corresponds to S: $0 + 0 = 1 - 1$. Since the strangeness conserve in this process, this process can be done by the strong interaction.

Then, at the decays, if neither C nor D have a choice to decay to other strange particles, they can not decay strongly. This makes the lifes of both C and D longer, and the lifes are long enough to make C and D to travel macroscopic distances. Then they decay to ordinary particles, E, F, G, H, as C → E + F, D → G + H. As a result, these decays are seen as the V-shapes.

This kind of events, observed in the cosmic-rays, can be interpreted as the following reactions using the modern notations.

$$\pi^- + p \to K^0 + \Lambda^0$$

where both π^- and p have $S = 0$, and K^0 has $S = 1$, Λ has $S = -1$. Then, each

$$K^0 \rightarrow \pi^| + \pi^-$$

and

$$\Lambda^0 \rightarrow p + \pi^-$$

makes the V-shape.

6.9 SU(3) and the Sakata model

Since the strange particles can not be made from non-strange particles, a member of the strange particles has to be added to a set of the fundamental particles. S. Sakata extended the isospin model of Fermi and Yang for the strange particles. The isospin model uses SU(2) because the isospin has two components: up and down. The extended model by Sakata, the **Sakata model**, which uses **SU(3)** to include strange particles. The SU(3) has three components: isospin-up, isospin-down and strangeness. He picked up proton, neutron and Λ as the three fundamental particles. Then mesons are constructed as listed below.

meson	content
π^+	$(p)(\bar{n})$
π^0	$(p)(\bar{p})$ and $(n)(\bar{n})$
π^-	$(n)(\bar{p})$
K^+	$(p)(\bar{\Lambda})$
K^0	$(n)(\bar{\Lambda})$
$\bar{K}^0$	$(\Lambda)(\bar{n})$
K^-	$(\Lambda)(\bar{p})$

In the above table, the bars above particle symbols indicate anti-particles.

Since the Sakata model is an extension of the Fermi and Yang model, it also has the problem of their model, unusually large binding energy for mesons.

6.10 Quarks

Gell-Mann proposed a model that the nucleon itself is a composite particle, made of three fundamental components. He named those fundamental components as **quarks**. G. Zweig also came to the same idea independently. The three types of quarks are named as up (u), down (d), and strange (s), using the today's naming scheme. Quarks are bonded inside the nucleon with the strong interaction. The particles, made of three quarks, are called **baryons**. The nucleons are baryons, so as Λ. Quark compositions of the typical baryons are summarized in the following table.

baryon	quark composition
p	$(u)(u)(d)$
n	$(u)(d)(d)$
Λ	$(u)(d)(s)$
Ξ^0	$(u)(s)(s)$
Ξ^-	$(d)(s)(s)$
Σ^+	$(u)(u)(s)$
Σ^0	$(u)(d)(s)$
Σ^-	$(d)(d)(s)$

There are many baryons other than the quoted ones in this table.

Mesons are made of a quark and an anti-quark. The following table lists the quark compositions of the typical mesons.

meson	quark composition
π^+	$(u)(\bar{d})$
π^0	$(u)(\bar{u})$ and $(d)(\bar{d})$
π^-	$(d)(\bar{u})$
K^+	$(u)(s)$
K^0	$(d)(\bar{s})$
K^-	$(s)(\bar{u})$
$\bar{K}^0$	$(s)(\bar{d})$
η	$(u)(\bar{u})$ and $(d)(\bar{d})$ and $(s)(\bar{s})$
η'	$(u)(\bar{u})$ and $(d)(\bar{d})$ and $(s)(\bar{s})$

The baryons and the mesons are together called as **hadrons**. The hadrons can interact strongly.

Though quarks were introduced in the early 1960s, they were not found experimentally in spite of many vigorous searches. So, many physicist thought that quarks are not real particles. But, after the discovery of so called J/ψ particle in 1974, the quarks are widely accepted as the real particles by the physicists. The J/ψ particle was nicely explained as a bound state of a charm quark and an anti-charm quark. Possible existence for the charm quark had been proposed since 1960s, but they were not widely accepted by the physicists until the discovery of J/ψ.

6.11   Classification of elementary particles

Two more types of quarks have been discovered since then. Today we know six types of quarks: **up** (u), **down**(d), **charm**(c), **strange**(s), **top**(t), **bottom**(b). Other than quarks, there are a group of particles that do not interact strongly, like electrons. Those are called **leptons**. We know six types of leptons: **electron** (e), **electron neutrino**(ν_e), **muon**(μ), **muon neutrino**(ν_ν), **tau**(τ), **tau neutrino**(ν_τ). In addition, there are a group of particles which work as mediating forces, like photons. And finally, the higgs particle was found in 2012. The table, shown below, summarizes those particles.

Matter particles.

division	type	1st generation	2nd generation	3rd generation
quark	up-types	u	c	t
	down-types	d	s	b
lepton	charged	e	μ	τ
	neutral	ν_e	ν_μ	ν_τ

where the word: **generation** means a family. There are three families, or generations. Each generation has one up-type quark, one down-type quark, one charged lepton and one neutral lepton.

Particles to mediate forces (gauge particles).

name	mediating force
γ	electromagnetic force
g(gluon)	strong force
W^+, W^-	weak force
Z^0	weak force

Higgs particle.

name
H

The matter particles obey the rule based on the Pauli's exclusion principle. Both the (gauge) particles, which mediate forces, and the higgs particle, do not obey the Pauli's exclusion principle. We'll describe the gauge particles and the higgs particles later.

Chapter 7

Imaging a Proton

It's interesting to look at the proton using an electron-microscope, like looking at a virus. The physicists in California did this in the 1960s using a very powerful, i.e. very high-energy, electron-microscope — "one mile electron linac."

7.1 Principle

When a high-energy electron hits a proton, it is deflected by the electric field of the proton. We call this phenomenon as a scattering of an electron by a proton, shown below.

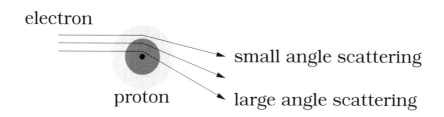

When the electron is scattered near the edge of the proton, it's scattered with a small deflection angle. When the electron is scattered by the proton's core, the deflection angle will be large. By analyzing the variations in the scattered angles of the electrons, we can examine the proton's electric field distribution — or the proton's structure.

To measure the proton's structure with a higher resolution, we need higher energy electrons. The reason for this is as followed. A scattering of electron is done by the force between the electron and the proton. In the other words, the electron receives a kick from the proton. The direction of this kick is approximately perpendicular to the direction of the incident electron. Then, the momentum by this kick is added to the incident electron's momentum, and the combined momentum is the scattered electron's momentum. Now, the Heisenberg uncertainty principle shows up.

(uncertainty in the transverse direction) × (transverse momentum)
= (h : constant)

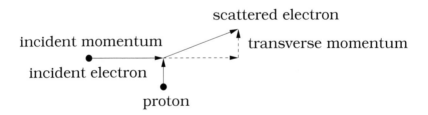

To improve the spacial resolution, the uncertainty in the transverse direction has to be made smaller. To do this, the transverse momentum has to be increased, because of the uncertainty principle. As shown in the above figure, a triangle is formed by the three momentum vectors: the incident electron, transverse momentum from the proton and the scattered electron. Since, to increase the transverse momentum, i.e height of the triangle, the momentum of the incident electron, i.e. baseline of the triangle, has to be increased, in addition to increase the angle of the triangle. Hence, we need higher energy electrons.

The devices to increase the electrons' energy are called **accelerators**. To look at the smaller structure of the proton, we need higher energy electrons because of the Heisenberg uncertainty principle. To make higher energy electrons, we need a larger accelerator. So, it is the Heisenberg uncertainty principle that makes the accelerator larger to look at smaller distance.

$$(\text{size of the accelerator}) \propto 1/(\text{resolution})$$

7.2 Inelastic scattering

To look at the proton by the electron-microscope, the proton should keep its shape, i.e. the scattering should be **elastic scattering**, as shown below.

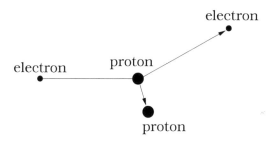

When electron's energy was increased for closer looks, however, the appearances of the scatterings changed. For higher energies, **inelastic scatterings**, as shown below, became dominated.

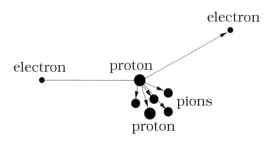

That is, many pions were produced at the scatterings. This means that the sample to examine (= proton) was broken. This should not happen at the initial thought. However, this inelastic scattering gave a surprising discovery.

7.3   Parton model

A proton can be expressed as a **wave packet**. A wave packet is made of a certain combination of waves. Actually, any wave can be decomposed to a mixture of pure waves of various wavelengths, which is called the "Fourier decomposition." The figure, shown below, illustrates a proton as a wave packet and its decomposition to various wavelengths of pure waves.

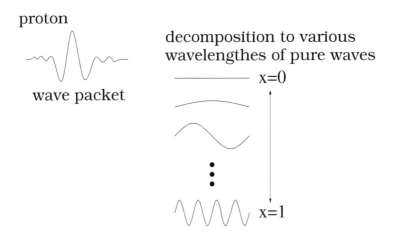

As shown above, the proton is made of a specific spectrum of the waves of various wavelengths, from the long wave, $x = 0$, to the short wave, $x = 1$. Those pure waves are called **partial waves**. So, a proton is made of a set of partial waves. From the wave-particle duality in the quantum mechanical system, the partial waves are also regarded as **partial particles**. Feynman called these objects as **partons**. Then he made a model to describe the proton using the partons, the **parton model**.

In the parton model, a proton is considered to be made of a set of partons. The partons are point particles. They are freely moving around in the proton. There's no interaction between partons. Partons are free particles as long as they are inside the proton. But they can not get out from the proton freely.

In the inelastic scattering process of the electron and the proton, the electron collides with one of the partons in the proton as illustrated below.

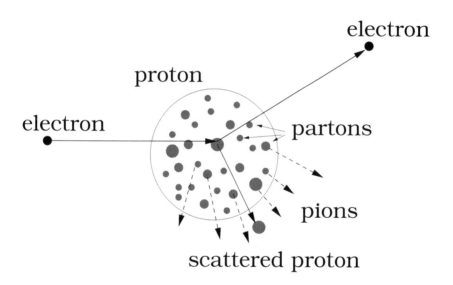

The scattering of the electron with the parton is an "elastic scattering." The scattered parton passes through the other partons in the proton without interaction. When the scattered parton exits the proton, however, it interacts strongly with the other partons. By this strong interaction, both the scattered parton and the other partons produce mesons, mainly pions. The proton itself is managed to recombine as a proton after these processes. As a result, the final state consists of many pions, the scattered proton and the scattered electron. This is what happens in the inelastic scattering.

A crucial point in the parton model is that the mass of the scattered parton can be determined by measuring the kinematics of the incident and the scattered electrons only. No information of the scattered proton or produced pions is necessary. Because of this feature, it is very simple to calculate the scattered parton's mass; the scattering is actually "elastic." If we write the proton mass as m and the parton mass as m_X and if we define their ratio, x, as

$$x = \frac{m_X}{m},$$

this x can be determined by measuring the momentum of both the incident and the scattered electrons. Since the proton should have a fixed mixture of partons, x distribution should be constant. The distribution should be independent of the incident momentum or the scattered momentum as long as their combination gives the same x. This phenomenon is called the **Bjorken scaling** and this x is called **Bjorken x**. The Bjorken scaling was first discovered by analyzing the experimental data. The meaning of this was not very clear at the beginning. But the parton model nicely explained the meaning of the Bjorken scaling.

This success supported the correctness of the parton model. Since the partons are freely moving inside the proton, the strong interaction seems to be weak at short distances. Since the partons can not escape from the proton without producing pions, the strong interaction seems to be very strong at long distances. This behavior, weak at short distances and strong at long distances, is called **asymptotic freedom**.

As we have seen in this chapter, the electron-proton (ep) scattering experiments changed their behavior from the elastic scattering to the inelastic scattering as the electron's energy increased. The inelastic scattering was not originally intended. It was rather annoying. However, this inelastic scattering made a clue to solve the mystery of the strong interaction, which will be explained later in this book. The partons, observed in the electron-proton scattering, are now considered as quarks (or anti-quarks).

Chapter 8

Electromagnetic Interaction

Since electromagnetic interaction is the most familiar interaction of all, we start the series of describing interactions from this interaction. Basic objects in the electromagnetic interactions are electric charge and electromagnetic field. For the electric charge, we simply call it "charge." In addition, we assume that electrons are the only charged particles in this chapter.

8.1 Gauge theory

We will explain the electromagnetic interactions using the concept of the **gauge theory**.

The charge of an electron, e, has a fixed value. All the electrons have the same charge, e. The charge stays constant in time, probably forever. This experimental fact is called the **conservation of electrical charge**.

Microscopic phenomena always have the wave-particle duality. They oscillate. According to this, all the quantities of electron oscillate. Their values are spread due to the Heisenberg uncertainty principle. Since charge is one of the quantities that the electrons have, it is no exception. It has to oscillate and its value has to spread. But in the reality, the electron's charge is a constant. It does not vary. The Heisenberg uncertainty principle tells that

(width of the canonical conjugate of e) × (variation of e)
$= (h : $ constant$)$.

Because the "variation of e" is zero, the **width of the canonical conjugate of e has to be infinite**. What is "canonical conjugate?" The canonical conjugate of a quantity is a partner of the quantity in the oscillation phenomena, i.e. the Heisenberg uncertainty principle. A quantity and its canonical conjugate have a relation between the distribution's width and its slope, as we mentioned in the chapter of quantum mechanics. In this case, since the charge is constant, the slope of the charge is zero. As a result, the width of the canonical conjugate of charge is infinite. It can take any value. In the other words, the nature is symmetrical under the change of this quantity. This quantity, the canonical conjugate of charge, is called **gauge function, χ**. The gauge function may not be the best name for its role, but it has been used for long time and the name is widely accepted. Historically, H. Weyl discovered it as a symmetry which connects the Maxwell equation and the electron, for which he named "gauge function."

Since electrons are described by the quantum mechanics, they have an ambiguity in their phase, given by the angle: $e\chi$, as shown below.

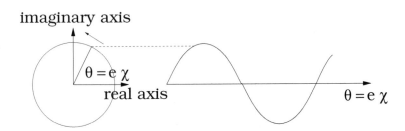

This is the standard behavior of quantum mechanical systems. One can change the phase of the system without changing the physical outcome. Then, we introduce an assumption. We assume that the physical system does not change even if the phase change has a dependency on its location: the location of the electron.

$$\theta = e\chi \rightarrow \theta(x) = e\chi(x)$$

In the above, θ means a phase angle that is constant in location, while $\theta(x)$ means a position dependent phase angle. Position depending variables are generally called as **local variables**. In this case, we assume that the gauge function, $\chi(x)$ is local.

What happens if the phase angle changes at location by location. We will explain this using a model with a rotating bar. If there's a bar, with a mass at its one end, which can rotate freely around a pole attached to the other end, as shown below.

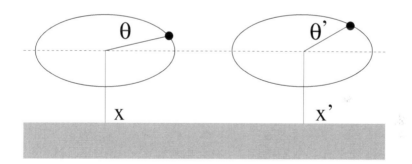

If the rotation plane is horizontal, any rotation angle is equally possible. In the other words, the system is symmetric under the rotation, or the system has a rotational symmetry.

If the rotation angle depends on the location, i.e. the angle is different at each position, as shown above. In this case, when we move the system from one place, x, (left side) to another, x', (right side), the angle of the bar has to change. This forces the system to rotate. Since there's a mass at the end of the bar, this rotation generates a centrifugal force to the bar. Of cause, this movement also generates an inertial force by the translation itself, but we ignore this effect for simplicity. If the bar is made of an elastic material and it can change its length, the length of the bar changes by moving from x to x'. This means that even the system has a rotational symmetry, the system loses its symmetry by moving, due to the effect of the forced rotating motion.

In the electrons case, the rotation angle of this system corresponds to the phase angle, $e\chi(x)$. When the electron moves, the movement forces the electron to change the phase angle, $e\chi(x)$. This rotation of the phase angle produces a kind of inertial effect and the state of the electron changes. If the electron's state changes by moving its location, we can not move the electron arbitrarily without avoiding the conflicts on the rotational symmetry. One may say that this problem is caused by the bad assumption. To start with, it was wrong to assume the position dependency in $\chi(x)$. But, there is a solution to solve this problem without discarding the assumption. The solution is, if the surrounding space has an ability to provide a canceling effect to this deformation, the electron can stay the same by moving its location. Actually, the space has such an ability, which is provided by the electromagnetic field. By the electromagnetic interaction between the electron and the electromagnetic field, the electron can keep the local phase rotation symmetry. We call this mechanism as **local gauge symmetry**.

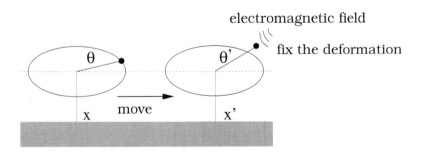

To make this trick clearer, we give an another explanation for this mechanism. If there is no electromagnetic field, the energy of an electron vastly increases when the electron moves, due to the forced phase rotation associated with the displacement. This large increase of energy requires a big work for a tiny movement. Since the work in physics is a product of force and distance, the combination of the tiny movement and the big work means that the force is huge to move the electron. Because of this huge force, the electron is locked at its location in vacuum. No electron can move in this situation. The existence of electromagnetic field, however, can make this locked electron move freely.

In addition to this, the electromagnetic field associates photons due to its wave-particle duality. So we can say that to keep the electron's local gauge symmetry, photons have to exist. The particle to provide the ability of keeping the local gauge symmetry is called **gauge particle**. A photon is a gauge particle. And the interaction to keep the local gauge symmetry is called **gauge interaction**. In this case, the electromagnetic interaction is the gauge interaction.

To summarize the arguments,

- The charge of the electron, e, is constant.

- All the characteristics of the electron have to oscillate.

- Since the charge, e, is one of the characteristics of the electron, it has to oscillate. This oscillation follows the Heisenberg uncertainty principle: $(e)(\chi) = (h$: constant$)$.

- In the Heisenberg uncertainty principle, since the gauge function, χ, is arbitrary, i.e. the gauge symmetry, the charge, e, does not change, or vice verse.

- If the χ changes at location by location, the state of electron changes by moving the electron.

- If the surrounding space has an ability to cancel this effect, the electron regain its ability of the gauge symmetry for its motion in the space.

- This ability of the space is provided by the electromagnetic field and the electromagnetic interaction between the field and the electron. In the other words, without electromagnetic field, electrons are locked inside the solid vacuum and no electron can move. The existence of electromagnetic field make it possible for the electrons to move freely in vacuum.

The theory to explain this relation between the electromagnetic field and the electron is called **gauge theory**. The gauge theory has been around for several decades. In the first few decades after the discovery, however, many physicists doubted its values. Since the gauge was not an observable quantity and the assumption of the locality of gauge transformation seemed to be a bit artificial, physicists thought it is interesting but it may be a toy. However, after the success of the Weinberg and Salam theory around 1970, few physicists doubt its value now. It is now commonly accepted as a major part in the standard model. We'll come back to the electromagnetic interaction in the next chapter.

Chapter 9

Strong Interaction

9.1 History

The strong interaction was found rather recently in 1933, mainly because no strong interaction is directly seen in our daily life. After the
discovery of the nuclei by E. Rutherford in 1911 and the discovery of
the neutrons by J. Chadwick in 1931, physicists were puzzled why the
positively charged protons and the electrically neutral neutrons get
together to form a nucleus. From the analogy of the electromagnetic
forces, H. Yukawa proposed a new force that attracts the protons and
the neutrons in the nuclei. He introduced a massive particle which
mediates this force. The mass of this particle was estimated from its
mediating range. Using the Heisenberg uncertainty principle, the mass
and the range have a relation

$$(\text{mass}) \times (\text{range}) = h.$$

Since this force was not seen in the daily life, he thought that its
range must be short, probably it only works inside the nuclei. By

substituting the typical size of a nucleus, 1 fm, in the above equation, the mass was estimated to be 200 times the mass of the electron. This particle was named as "meson" because its mass is in between the electron and the nucleon.

The strength of this force has to be much stronger than the electromagnetic force since it has to outdo the repulsive forces by the electromagnetic interactions between the protons. Because of this strength, the new force was named as **strong force**. And the interaction that provides this force was called the **strong interaction**.

How strong is the strong force? To estimate this, we consider the stability of heavy nuclei. All the nuclei heavier than the lead nucleus are unstable. The electromagnetic forces of all protons work cooperatively. So the repulsive forces of all the protons in the nucleus accumulate. Since the lead nucleus has 82 protons, the strength of the net repulsive force is 82 times the strength of a single proton. The strong force, on the other hand, works individually. So, by combining the electromagnetic forces of 82 protons, it can surpass the attraction by the strong force. Because of this, the nuclei, heavier than the lead nucleus, disintegrate. From this simple arithmetic, the strength of the strong force is estimated to be about 100 times stronger than the electromagnetic force.

After the discovery of the meson, the strong interactions were studied vigorously by many experiments. A huge amount of experimental data were accumulated in the 1950s and 1960s. However, the strong force had been kept concealed for the long time. It was only in the early 1970s when the underlying mechanism of the strong interaction was finally revealed. It came out from an unexpected place.

9.2   Color

When classifying the elementary particles using the quark model, there were problems found during the 1960s. An example for the problems is

Δ^{++}, a well known baryon. Typically Δ^{++} is produced by colliding a pion to a proton. It immediately decays to a pion and a proton. From these production and decay modes as well as its mass, Δ^{++} is likely to be made of three up-quarks. This, however, conflicts the Pauli's exclusion principle. Matter particles of the same species, such as up-quarks, with the same property, can not be put together in a tightly confined place due to the Pauli's exclusion principle. In this case, up to two up-quarks can be put together since one up-quark can be spin-up state and the other up-quark can be spin-down state. But, there's no way to put all three up-quarks inside Δ^{++} since the up-quark's spin can be only up or down.

To overcome this difficulty, there were many proposals. Among them was color hypotheses. If we assume that the quarks have a new property, which was not known at that time, and if the three up-quarks inside Δ^{++} are different in that property, they can be put inside Δ^{++}. If anything is different, we can avoid the Pauli's exclusion principle. The new property was named **color**.

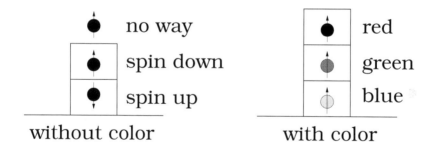

The color of a quark is completely different from the color in our daily life. It is nothing to do with the wavelength of light. But the quark's color has a similarity with the light's color. There are three quark colors: red, green and blue. The reason for the three is to distinguish the three quarks inside the baryon, such as Δ^{++}. In addition, if we assume that all physical states have to be white color, or colorless, then we can explain why there are only baryons with three quarks and mesons with one quark and one antiquark naturally exist in our world.

If red, green and blue are mixed, we'll get white, giving a baryon. If red and anti-red (which is cyan) is mixed, we'll get white, giving a meson. On the other hand, if red and green are mixed, for example, we'll get yellow, not white. So there is no such particle exists. Up to the combinations of three quarks or antiquarks, only the combinations of three different colors, or the color and its anti-color, can make white. There is no diquark, which is a combination of two quarks, because the mixtures of two colors can not be white.

As an example to show the color-flows in the strong interaction, a Feynman diagram with color-flows is shown below.

In the above figure, two protons, each with three colors, interact by exchanging a pion, made of red and anti-red. The red of one proton flows through the pion to the another proton. The red of the another proton flows back to the proton through the pion. These flows of colors keep the colors of all particles white.

As seen above, the quantity, color, was introduced to avoid the difficulty in the quark model. But, this color gave us a key to solve the mystery of the strong interaction.

9.3 Color SU(3)

Since all the observable particles have to be white, we can not directly observe the color. There's no way to tell a pion is made of red and anti-red, or green and anti-green, for example. There's no difference in strong interactions caused by the difference in color. No observable quantity changes when the color is changed. These characteristics of the color is expressed that the physics is symmetric under the change of color, or the physics has **color symmetry**.

Though there's no way to detect color, there are rules on color, such as three colors exist, or only the white particles can be observed. So, there must be an underlying structure in the color symmetry. What kind of structure does the color have? Since there are three colors, the symmetry could be 3D rotation. Or, it could be 2D rotations with 120° angles. It turned out that the appropriate symmetry is **SU(3)**, which is used to classify the u, d, s quarks. Why it's always SU(3)? The reasons are, firstly, we are dealing with the quantum mechanical systems which always have unitary symmetries, and secondly, there are three colors — 3 in SU(3) comes from three colors. We'll explain what characteristics the SU(3) color symmetry will bring us.

In the chapter for the particle classification, we mentioned that 2D rotations form a group. Three dimensional rotations also form a group as shown below.

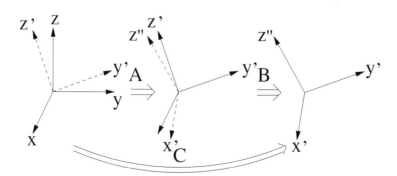

In the above figure, first the left system is rotated around the x-axis. After this rotation, the system will be transformed to the center system. We call this operation A. Next, the center system is rotated around the y-axis. After this rotation, the system will be transformed to the right system. We call this operation B. If we select a proper rotation axis and a proper rotation angle, we can directly rotate the left system to the right system by a single operation, denoted as C. Then the operations, A, B, C have a relation.

$$A + B = C$$

From this result, we can say that A, B, C form a group. This group is the 3D rotation group.

There is a critical difference between the 2D rotation group and the 3D rotation group. In the 2D rotation group, if we flip the order of two operations, the result is the same. For example, if the system is rotated by $30°$ clockwise (operation: D), followed by a rotation of $20°$ in the counter clockwise direction (operation: E), the result is a rotation of $10°$ in the clockwise direction. If we swap the order of the operations, they will give us the same result. So we can write

$$D + E = E + D$$

When the order of operations does not change the result, we call such operations **commutative**. Two dimensional rotations are commutative. However, 3D rotations are not commutative. As shown in the figure below, if we change the order of 3D rotations, the result will be different. In the figure (a) below, the system is rotated around the x-axis (operation: A), followed by the rotation around the y-axis (operation: B), the result is shown at the top-right. In the figure (b) below, the order of the operations A and B are swapped. The result is shown at the bottom-right.

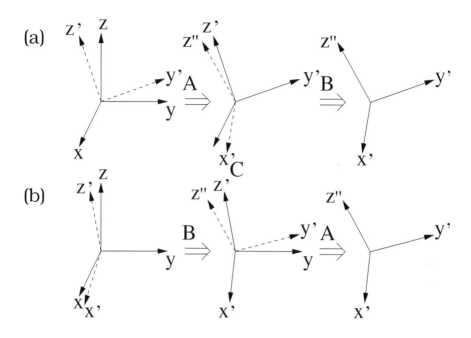

Though it is not clear from the above figure, the two results are slightly different. This is written as

$$A + B \neq B + A.$$

If the operations of two members of a group can be swapped, the group is called **abelian group**. If they can not be swapped, the group is called **non-abelian group**. The 2D rotation group is an abelian group, while the 3D rotation group is a non-abelian group.

Previously in the particle classification chapter, we mentioned that SU(2) has three degrees of freedom, thus, it behaves similarly to the 3D rotations. Hence, SU(2) is a non-abelian group. Since SU(3) is an extension of SU(2), SU(3) is also a non-abelian group. SU(3) has eight $(8 = 3^2 - 1)$ degrees of freedom, thus, it behaves similarly to 8D rotations.

In the electromagnetic interactions, we explained that the conservation of electron's electric charge implies the gauge symmetry, or vice verse. Though color is different from the electric charge, it behaves like a charge, thus, we call it **color charge**. All the quarks are charged up with color. Since color charges are quark's physical quantity, it can oscillate in principle. Hence, it has to obey the Heisenberg uncertainty principle,

$$\text{(gauge function of color)} \times \text{(color charge)} = (h : \text{constant}).$$

Since color charges are constant and they do not change, the amplitudes of their oscillations are zero. Thus, the gauge function, corresponding to the color, can vary infinitely. Because of this, the color gauge function can change freely without modifying the strong interactions. In the other words, strong interactions have **color gauge symmetry**.

Comparing the color charge with the electric charge, the electric charge can be classified by a sign, usually denoted as positive or negative after B. Franklin. The color charge has three types, or three colors. The underlying symmetry of these three colors are SU(3) symmetry. Thus, the corresponding color gauge functions have to have the SU(3) symmetry too. This situation is similar to the case of position vector, x, that had the corresponding momentum vector, p.

Then, we assume that the color gauge functions are local, i.e. they change at location by location, similarly as the electromagnetic gauge functions. This position dependency generates a force to rotate the quark state by moving the quark, producing an inertial force and a deformation caused by the force. To cancel this deformation, there has to have a field which acts on the quark to keep its original form. These mechanisms are same as the electromagnetic case. In the strong interaction case, or the quark's case, the field to work is called **gluon field**. The particles, associated to the gluon field, are called as **gluons**. The gluons correspond to the photons in the electromagnetic field. The interaction, introduced in the above, is the strong interaction. It is the interaction between the quark and the gluon.

As we have seen, the mechanism of strong interaction has one-to-one correspondence to the electromagnetic interaction. A big difference, however, comes from the difference in the underlying symmetries, in addition to their strengths. The strong interaction has color symmetry, which is SU(3). And SU(3) is a non-abelian, i.e. non-commutative. The symmetry in the electromagnetic interaction is simpler. It is called U(1). And U(1) is an abelian, i.e. commutative. This difference in commutability causes the big difference in their behavior. Commutable means no interference between operations. In the words of electromagnetic interaction, it means there's no interaction of electromagnetic field by itself. In the words of elementary particles, this means that a photon does not interact with other photons. In the case of the gluon field, this is not true. Since SU(3) is non-commutative, there are interferences between operations. It is non-commutative because of these interferences. This means that there are interactions of gluon field by itself and the **gluons interact with other gluons**.

An interaction between a quark and a gluon is shown below.

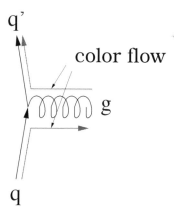

In the above Feynman diagram, the spring like line shows the gluon. The arrows show the quark. A color flows from the initial quark to the gluon, and an another color flows from the gluon to the final quark.

Because of these color flows, the quark changes its color by the interaction with the gluon. Since two colors simultaneously flow on the gluon line in the opposite directions, gluons are made of a color and an anti-color. Contrarily to mesons, gluon's color and anti-color are not complementary. They do not cancel each other. As a result, gluons are not white, hence they can not be observed directly.

Since gluons have color, they interact with each other. This agrees with the non-commutative nature of the gluon field. The figure, shown below, shows Feynman diagrams of gluon-to-gluon interactions. The gluons are indicated by g.

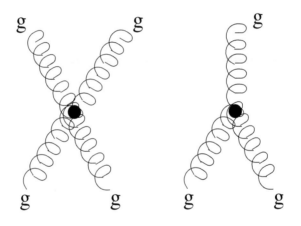

Correspondences between the electromagnetic interactions and the strong interactions are listed below.

electromagnetic interaction	strong interaction
electron ↔ photon	quark ↔ gluon
electric charge (1 type)	color charge (3 types)
U(1)	SU(3)
commute	non-commute
no photon-to-photon interaction	gluon interacts with other gluon
photon has no electric charge	gluon has color charge

U(1) in the above table means one dimensional unitary group. This is actually a simple phase rotation, i.e. a 2D rotation.

To summarize the story up to here, quarks have three colors, and the colors conserve. This conservation requires the existence of gluon field. Because of the non-commutative nature of the colors, gluons interact with other gluons.

9.4   Quantum chromodynamics

The interactions of quarks and gluons via the colors are described by **Quantum ChromoDynamics (QCD)**. QCD was developed in the early 1970s by many physicists. It had been a long time since Yukawa first proposed the meson theory to explain the strong interactions. The attractions between the nucleons in a nucleus were finally understood as a residual force of quark-gluon interactions. The reason, why a colorless particle such as proton or neutron interact strongly, is due to the mechanism similar to the attractions in the molecules. Molecules are made of atoms by the attractive forces caused by the electromagnetic interactions. The atoms, however, are electrically neutral objects. So why electrically neutral objects, like atoms, attract each other by the electromagnetic interactions? As you had learned in the chemistry class, they use covalent bonds, coordinate bonds, etc. to get together. They are basically residual forces of electromagnetic interactions. The nuclear forces are similar to this. They are strong interactions but they are not the fundamental interactions.

Next, we compare the structures of an electron and a quark. The differences in the structures come mainly from the differences between the photons and the gluons. Photons do not interact each other. This is equivalent to say that photons do not have electrical charge. Gluons do interact each other, or one should say that gluons have color charges. The figures, shown below, illustrate their structures.

electron's structure   quark's structure

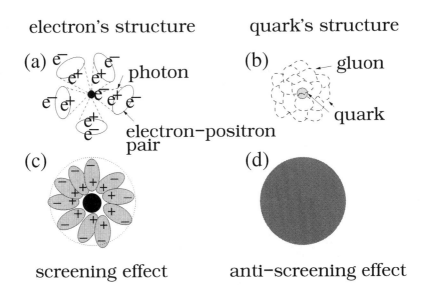

screening effect   anti-screening effect

As shown in the figure (a), if there is an electron, the electron is surrounded by photons. Those photons are called virtual photons because they only exist for a very short time. These virtual photons produce electron-positron pairs. This sounds a violation of energy conservation. This really is. But in a very short time interval, energy conservation can violate. Violation of the energy conservation is allowed as long as the magnitude of violation and the time interval satisfies the Heisenberg uncertainty principle. Once the electron-positron pairs are made, they apart each other in spite of their opposite charges. This is due to the repulsive forces between them, caused by the Pauli's exclusion principle. In addition, the positrons in the pairs are attracted by the original electron. And the electrons in the pairs are repulsed by the original electron. As a result, polarization appears around the original electron, as shown in the figure (c). This cancels a part of the original electron's charge. This phenomenon is called **screening effect** of the electron's charge. Because of this screening effect, the effective electrical charge of the original electron is reduced. So when we look at the electron from a distant point, the electric charge is seen to be small. But when we get close to the electron, the electric charge looks large. In the other words, electromagnetic interaction is weak at a long distance but it is strong at a short distance.

This phenomenon can be also explained by the hole theory. In the hole theory, the electron will make the surrounding negative energy electrons to move away, making holes around the original electron. These holes will shield the original electron's charge, which is the screening effect, mentioned above.

Quarks, on the other hand, have completely different structures. As shown in the figure (b), a virtual gluon, emitted from the original quark, can produce other gluons. This can only happen to gluons, not photons, since the gluons have color charges, hence they interact with other gluons. Photons, on the other hand, do not have electrical charge and they can not produce other photons. Because of this gluon-makes-gluon effect, the original quark is surrounded by sea of gluons. In addition, gluons are free from the Pauli's exclusion principle. They are not repulsive with each other. Actually gluons attract other gluons, making a gluon nugget. Since gluons have color charges, they enhance the original quark's color. This effect is called **anti-screening effect**, as shown in the figure (d). The gluons also produce quark-antiquark pairs which behave similarly as electron-positron pairs in the electron case. However, since the number of quark types are not large compared to the number of gluons, which is eight, the gluon's effect overcomes the quark-antiquark pair's effect. As a result, the original quark's color is spread out and it looks like a nugget of same color. Because of this mechanism, when we look at a quark at a long distance, we can see a strong color. However, when we get close to the quark, the color fades away, because the effect of the surrounding gluons' color is gone. This is similar to the case of the cloud in the sky. When we look at a cloud at a long distance, it has white color. But, when we get into the cloud, it is not as white as it used to be. It is almost transparent. Strong interactions are strong at long distances, but they are not that strong at short distances. This agrees with the **asymptotic freedom** phenomena, observed in the electron-proton scatterings.

Charge distributions, for the electron and for the quark, as a function of the distance from the each center are shown below. For the electrons, the charge distribution has a peak at zero distance, $r = 0$. The charge density decreases rapidly as the distance increases, due to the screening effect. For the quarks, the color charge has a flat distribution. There's no clear center seen in the distribution.

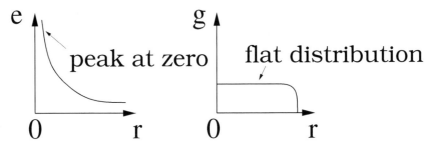

This behavior of sharp increasing of the strength at short distances in the electromagnetic interaction is called **ultra-violet divergence** using the analogy of light wave. The ultra-violet means short wavelength, that is high energy, or high momentum. And the high momentum means short distance. The divergence is a word for a rapid increase of its strength. The behavior of the strong interaction at the short distances is opposite to this, as it is called the asymptotic freedom.

9.5   Why we can not find quarks?

The quark model can classify the hadrons successfully. Quarks and gluons play the essential role in QCD, which can explain the strong interaction successfully. After the discovery of J/ψ particles in 1974, few physicists doubt the existence of quarks. But we still have not found the quark yet. Why we have failed to find quarks? Physicists now believe that quarks can not be found. Quarks are colored objects which can not escape from the hadrons. In addition, we now know the mechanism that the quarks can not come out from the hadrons. This mechanism is called **quark confinement**.

The figures, shown below, illustrate the electromagnetic field for a pair of positive and negative charge (left) and the gluon field for a pair of quark-antiquark (right).

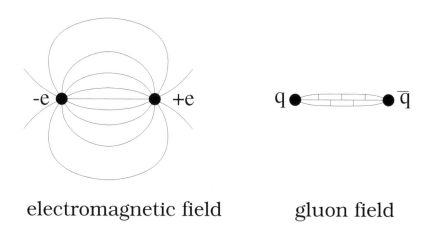

electromagnetic field   **gluon field**

In the electromagnetic field case, the electric field lines flow from the positive charge to the negative charge. Accompanying to these field lines, virtual electron-positron pairs are created which are repulsive due to the Pauli's exclusion principle. As a result, electric field lines are repulsive with each other. This makes the field lines widely extended in 3D, as shown in the figure (left). If we pull one of the charges away from the other. The field lines will extend even wider in the 3D space. As a result, the attractive force between the positive and the negative charges gets weaker with the distance. Hence, we can extract the charge as a free charge.

In the gluon field case, on the other hand, virtual gluons are created on the gluon field lines. Gluons are free from the Pauli's exclusion principle. Moreover, gluons are attractive with each other. Quark-antiquark pairs are also created on the gluon field lines. But the attractions by the gluons overcome the repulsions by the quark-antiquark pairs be-

cause of the surplus of gluons in the number. So the gluon field lines
are pulled together and make a bundle, or a string, as shown in the
figure (right). This string works similar to a rubber string. If we
pull one of the quarks away from the other, the gluon string extends,
making the attractive force stronger. This is opposite to the electromagnetic case. If we pull the quark further, the potential energy in the
gluon string increases further and the string eventually breaks. When
it breaks, a pair of new quark-antiquark are created at the each end
of the string. This is similar to what happens when we try to extract
a magnetic charge from a bar magnet. For a bar magnet, if we try to
separate N-pole from S-pole by breaking the bar in the middle, we will
get two bar magnets instead of the one with N-pole and the other with
S-pole. This also happens to the quark-antiquark pair. As a result of
breaking the quark-antiquark pair, we get two sets of quark-antiquark
pairs, as shown in the figure below, and we fail to extract the quark
as a free particle. Hence, we can not find quarks.

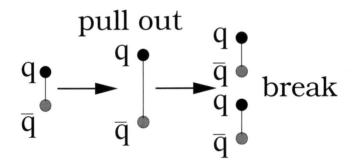

Chapter 10

Weak Interaction

The weak interaction works only in the nuclear scale and we rarely see it in our daily life. The most familiar phenomena of the weak interaction is probably radioactivities. The radioactivities are consequences of nuclear disintegration provoked by the weak interactions and other interactions. So we start with the nuclear disintegration.

10.1   Nuclear disintegration

Unstable nuclei change their contents by nuclear disintegrations. As a result of these disintegrations, radioactivities are the usual outcomes.

The radioactivity has three kinds: **α decays**, **β decays** and **γ decays**. Among these three, the α decays are caused by the strong interactions, the γ decays are caused by the electromagnetic interactions, and the β decays are the ones that the weak interactions concern.

The α decays are the phenomena that nuclei emit α particles. The α particle consists of two protons and two neutrons. The reason, why

they are in the units of α particles, is that the binding of this combination is very strong. In the other words, the binding energy of this
combination is exceptionally large. The figure, shown below, illustrates
a process of alpha decay.

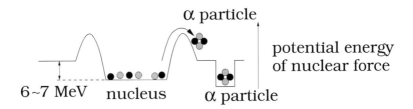

When two protons and two neutrons combine, they release the energy
of 28.3 MeV as

$$2p + 2n \rightarrow \alpha + 28.3 \text{ MeV}.$$

In the other words, the mass of an α particle is 28.3 MeV/c^2 lighter
than the total mass of two protons and two neutrons.

In addition to the α particles, nuclei also have binding energy. Because
of this, the potential energy of nuclear force in a nucleus is lower
than its outside, as shown in the above figure. In a stable nucleus,
the potential depth is $6 \sim 7$ MeV. In a heavy nucleus, the depth is
shallower. Hence, it will release energy if the nucleons escaped to the
outside of the nucleus as an α particle.

The nuclear potential of a nucleus, however, is surrounded by a barrier
as shown in the above figure. It seems that this barrier inhibits the
nucleons to escape. In the world of quantum mechanics, however, it is
possible to escape through this barrier by the Heisenberg uncertainty
principle. The α particle can get enough energy to go over the barrier
in a very short interval of time by oscillations. This phenomenon is
called **tunneling effect**.

The gamma (γ) decays are the phenomena that the nuclei emit γ rays, or high-energy photons. They usually take place after the α decays or β decays. After those decays, the nucleus sometimes get to a high potential energy state. In such a case, the nucleus is unstable and it emits a γ ray and it goes down to the stable state.

The beta (β) decays are the phenomena that either proton or neutron changes to the other type. The β decays can be classified into β^- and β^+. The β^- decay is a reaction that a neutron changes to a proton. In this reaction, an electron and an anti-electron-neutrino are also emitted as below.

$$n \to p + e^- + \bar{\nu}_e$$

The reason for this reaction to occur is that the neutron is 1.3 MeV/c^2 heavier than the proton. Because of this mass difference, neutrons are unstable. If a neutron is placed outside of the nucleus, it decays spontaneously with a life time of about 15 min. The reason, why the neutrons can be stable inside the nuclei, is due to the existence of nuclear force potential. As explained at the α decays, neutrons in the nuclei are in the potential well of nuclear force, it takes extra energy for the decay products to get out from the nuclei. If this potential is high, which is true in the stable nucleus, the energy conservation law prohibits this reaction. The β^- decays usually take place in the neutron rich nucleus. Such nucleus is heavy, compare to the nucleus with a balanced number of neutrons and protons. So it can save energy, if one of the neutrons decay to a proton.

The reason, why the neutron is heavier than the proton, is that the neutron has two d-quarks and one u-quark while the proton has one d-quark and two u-quarks. Then the point is that the d-quark is heavier than the u-quark. A part of this difference is canceled by the proton's electromagnetic energy. But this cancellation is not large enough. As a result, the neutron is heavier than the proton.

Inverse to the β^- decay is a β^+ decay, which is

$$p \rightarrow n + e^+ + \nu_e.$$

Since protons are lighter than neutrons, this reaction can not happen outside the nuclei. This reaction, however, can happen inside the proton rich nuclei. Protons have positive electrical charges, thus they repulse each other. If there are too many protons inside a nucleus, electromagnetic energies build up. Hence, it is more stable for the nucleus to change one of the protons to a neutron.

The positron, emitted by this reaction, will immediately hit an electron in the surrounding material. And both of them annihilate to two photons, or γ rays. Thus, in the β^+ decays, what we observe is a pair of γ rays, not positrons.

This mechanism is used in Positron Emission Tomography (PET). In the PET scan test, a patient is injected a kind of glucose with ^{18}F which is radioactive and decays to β^+. Those glucoses will be concentrated to energy guzzlers such as cancer cells. Then they decay, emitting β^+, thus emitting a pair of γ rays with a fixed energy, coming out in the opposite directions. By detecting those γ rays, we can reconstruct the image of the cancer.

In the accidents of nuclear reactors, a complex series of nuclear disintegrating processes will take place. One of the major fallout products is ^{137}Cs. This ^{137}Cs will β decay as

$$^{137}\text{Cs} \rightarrow ^{138}\text{Ba}^* + e^- + \bar{\nu}_e.$$

The half life of this reaction is about 30 years. The decay product of this reaction is ^{138}Ba* which is unstable and it immediately γ decays to the stable ^{138}Ba. This follow up decay emits a 662 MeV γ ray which has a strong penetrating power, causing most of the radiation problems.

The lifetimes of those two decays explain the weakness of β decays, comparing to the electromagnetic interaction. In the first decay: ^{137}Cs $\rightarrow$ ^{138}Ba*, the lifetime is about 30 years. In the second decay: ^{138}Ba* $\rightarrow$ ^{138}Ba, the lifetime is about 10^{-15} seconds. The first decay, β decay, is much slower than the second decay, γ decay. This explains that the weak interaction, which provokes β decays, is much weaker than the electromagnetic interaction, which provokes γ decays. The weakness of the weak interaction causes the slowness of the β decay. Because of this slowness, the radiation problems last long after the nuclear accidents.

10.2 Fermi theory

In the 1930s, it puzzled physicists that, why the energy of β rays are not constant, when the mass difference between the initial nucleus and the final nucleus is constant. They expected that, since the mass difference is constant, the emitted electron should have a constant energy, if the energy conservation law is valid.

To solve this energy crisis, W. Pauli proposed that a mysterious neutral particle must be escaped detection in the β decay. And this neutral particle takes the missing energy.

Though Pauli did not publish his idea, a young physicist, Fermi, used this neutral particle in his **theory of β decay**, published in 1933. In his publication, Fermi named this mysterious particle as **neutrino**.

The figure, shown below, indicates a Feynman diagram of β decay, using the Fermi theory.

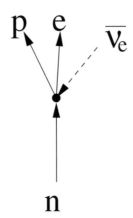

In the diagram, the neutron changes to a proton at a single point. When this change is taken place, the neutron also emits an electron and an anti-neutrino. This anti-neutrino is expressed as an arrow coming back from the future, then at the interaction point, this backwardly moving neutrino changes to the electron, and this electron moves to the future. In the figure, the overline at ν_e indicates that it is an antiparticle. At the time of the introduction of the Fermi theory, it was not known that the neutrinos have variations. The Fermi theory could explain the β decays very well.

10.3   Difficulty in the Fermi theory

In this section, we temporary change the subject.

Near the beginning of 1970s, at Batavia in Illinois, a new National Accelerator Laboratory, later named as Fermilab, was open. Using the new 500 GeV accelerator at the laboratory, quantitative measurements of high-energy neutrino interactions were possible at the first time.

As one of the results of the experiments, collision frequencies of neutrino-nucleon interactions were obtained. The figure, shown below, schematically shows the collision frequencies as a function of the incident neutrino's energy.

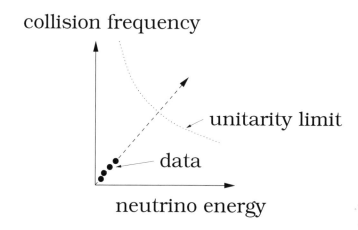

In the figure, the close circles (•) show the measured data. The horizontal axis shows the incident neutrino's energy. The vertical axis shows the collision frequency. The dashed line shows the fit to the data.

The measured data at that time showed that the collision frequency is proportional to the incident neutrino's energy. By extrapolating the fitted line to the data toward the higher energy, the collision frequency seems to increase linearly with energy.

The Fermi theory also supported this measurement. In the Fermi theory, the collision takes place at a single space point with a constant strength. As the incident neutrino's energy increases, the neutrino's energy in the center-of-mass system increases with a square root of the incident energy. And the number of possible scattering patterns increases with the square of the center-of-mass energy. Because the collision frequency is the product of interaction strength, which is constant, and the number of possible scattering patterns, which is proportional to the incident energy, the collision frequency should increase linearly with the incident energy.

In the above figure, a hyperbola-like dotted curve shows the unitarity limit. To explain this limit, first, the Fermi theory assumes that the interaction takes place at a single space point. This is true for neutrino-electron collisions since both the particles are point-like. Though the measured interactions are neutrino-nucleon collisions, we will limit the discussion to neutrino-electron collisions to make the situation simple. If the interaction is point-to-point, the distance between the two particles is zero at the collision. If the distance is zero, there can be no moment of rotation, thus the orbital angular momentum at the collision is always zero. In the other words, all the collisions are head-on collisions. Or there is no peripheral collision. This limits the frequency of collisions, which is the unitarity limit. So at least for neutrino-electron collisions, the collision frequencies have to be below this limit. On the other hand, the Fermi theory predicts the linear increase in the frequency. Thus, at one point, the linearly increasing collision frequency will cross this unitarity limit. This is certainly an internal conflict in the Fermi theory.

This problem was solved shortly after the above measurements. The Fermi theory was an approximate theory which works only at low energies. At high energies, the interaction takes place by a mediating particle, $W^{\pm}$, predicted by the Weinberg-Salam theory, which will be explained later in this chapter. In the Weinberg-Salam theory, since the interaction takes place not at a single space point, there is no conflict. The collision frequency does not increase linearly with energy.

This historical experience warns us that, if there is a conflict expected by the extrapolation in the energy, there must be a new physics beyond that energy. This is one reason why we do not think the currently successful standard theory is not the final theory. We believe that it must be replace by a new theory at higher energies.

10.4 Parity violation

When the space is inverted, laws of physics may or may not change. We call this behavior as **parity**. Space inversions are also called as **parity inversions**. The parity was discovered by E. Wigner as a property in the quantum mechanics in 1927.

The parity inversion is usually explained as a mirror image. When we look at the mirror, we see our face looking at us from the other end of the mirror. A mirror can invert the direction along the axis perpendicular to the mirror surface, as shown in the figure (a) below.

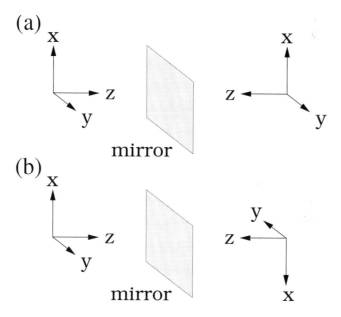

A mirror reflection inverts only the z-axis as shown in the above figure (a). It does not change the other directions, x and y. The true parity inversion inverts all three axes, as shown in the figure (b). We should be careful when we use a mirror to explain the parity inversion.

Winger found that the quantum mechanical system does not change

when we invert parity. This is called **parity invariance**. It had
been thought that all the physical systems have parity invariance after
Wigner.

Though the assumption of parity invariance seemed to be quite trivial
to the physicists, it was not true in the weak interaction of elementary
particles.

Around 1955, there were strange phenomena, observed in two kinds
of newly found particles. They were called τ^+ and θ^+. A τ^+ particle
decayed to three pions as

$$\tau^+ \to \pi^+ + \pi^- + \pi^+$$

while a θ^+ particle decayed to two pions as

$$\theta^+ \to \pi^+ + \pi^0.$$

The pions have odd parity, which was already known at that time. So
the system with three pions has a parity, P, as

$$P = (-1)^3 = -1$$

Hence τ^+ should have odd parity.

On the other hand, the system with two pions has a parity, P, as

$$P = (-1)^2 = +1$$

So θ^+ should have even parity. It should be noted that if a multiple
pion system has a non-zero orbital angular momentum among the

pions, the sign of the parity can change. But in the above systems, the orbital angular momentum is zero, thus they have no effect to the parity.

Since τ^+ and θ^+ have different parities, they should be different kinds of particles. However, the masses of τ^+ and θ^+ were same within the measurement errors. So the question of, why those two different particles have a same mass, puzzled the physicists.

T. D. Lee and C. N. Yang gave an answer to this question in 1956. They proposed that parity does not conserve in the weak interactions. Since τ^+ and θ^+ decay to multiple pions via weak interactions, the parity does not have to conserve in the decays, they thought. If this is true , τ^+ and θ^+ can be the same particle with different decay modes. This explains the equality in the mass of the "two" particles.

To find out the correctness of their proposal, C. S. Wu quickly pre-pared an experiment using β decays of ^{60}Co. She cooled down ^{60}Co to an ultra-low temperature. Then the ^{60}Co was placed in the strong magnetic field. And high frequency microwaves were irradiated to the ^{60}Co. The microwaves acted on the ^{60}Co's spin and they made the spin directions aligned with respect to the magnetic field. During the experiment, she flipped the direction of the magnetic field and mea-sured the number of electrons emitted by the β decays, using a detector located at a particular direction with respect to the magnetic field.

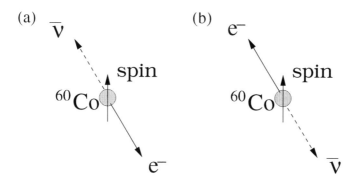

The above figure shows the schematics of what she measured. In the figure (a), the rate of electrons from the β decays was measured in the direction, opposite to the spin. Inverse to this, in the figure (b), the same rate was measured in the direction, same to the spin. She observed the difference in those two rates.

The (a) and (b) are in the parity inversion relation. In rotating motions, like the spin in this case, the direction of rotation does not change by the parity inversion. As shown in the figure below, if there is a rotating object A, which has the rotation center, indicated by the vertical arrow, and the distance a from the center, moving with a velocity c. If the parity of this system is inverted, A $\rightarrow$ B, $a \rightarrow b$, $c \rightarrow d$. As a result, we will get the same rotation. Hence the rotation is invariant under the parity inversion.

rotation axis

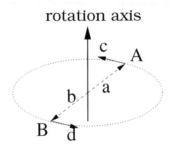

On the other hand, the moving directions of the electron and neutrino will flip by the parity inversion. As a result, if we invert parity in (a) in the above figure, we will get (b). So, if the parity is conserved, the rates of the emitted electrons should be the same for (a) and (b). But they were different. This proved that the **parity does not conserve** in the ^{60}Co's β decays, or more generally, in the weak interactions.

This discovery of parity violation in the weak interaction was very shocking to many physicists at that time. For this discovery, Lee and Yang received the Nobel prize in 1957. However, the Nobel committee did not give the prize to Wu.

10.5 Neutrino's helicity

Right after the discovery of the parity violation, M. Goldhaber measured the neutrino's helicity using a very clever method in 1958.

Helicity, H, is a variable that indicates the directing degrees of spin, measured along the motion of the particle. If the spin of a massless particle is directing at the direction of its motion, then $H = 1$. If the particle has a finite mass, then $0 < H < 1$. If the spin direction is opposite, then $H = -1$ for the massless particles, or $-1 < H < 0$ for the massive particles. The helicity is especially useful for the massless, or light mass, particles like neutrinos.

As similar to the helicity, there is a variable called **chirality**. The chirality is not a quantitative variable and it can take only two selections: **right hand** or **left hand**. As shown below, using a hand, the moving direction can be indicated by the thumb and the rotating direction is indicated by the other fingers. If the spin direction is the same as the moving direction, then it can be shown by the right hand. If the spin direction is opposite to the moving direction, then it can be shown by the left hand. For a massless particle, $H = 1$ corresponds to the right hand, $H = -1$ corresponds to the left hand.

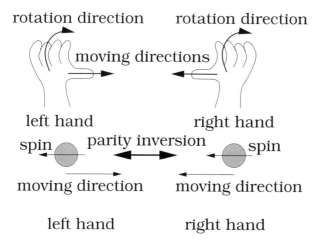

By a parity inversion, the direction of motion is reversed, while the spin direction is kept same, as shown in the above figure. So a parity inversion will flip the right hand and the left hand.

Goldhaber found that the neutrino's chirality is left hand, using a desktop experimental apparatus. It is amazing that the helicity of such an elusive particle as neutrino was measured by a small apparatus. It should be noted that the neutrino was first discovered by F. Reines and C. L. Cowan in 1956, just two years before the Goldhaber discovery, using a huge device: tons of liquid scintillator. Unfortunately, Goldhaber did not receive the Nobel prize (I think he should have received it.).

10.6   Theory for parity violation

The original Fermi theory is parity invariant. But the parity violation can be accommodated in this theory with a minor modification. Feynman and Gell-Mann did this. Feynman mentions the story of this discovery in his books. He was very much excited about his discovery, but he later found out that Gell-Mann also did it independently. M. Morita, in the Colombia University at that time, also found it independently, but he missed the chance for publishing it due to the internal restrictions in the university.

10.7   Discovery of *CP* violation

After the discovery of parity violation, invariance of all the symmetries were re-examined. One of the targets was the **CP symmetry**. What does the *CP* symmetry mean? *C* stands for **charge conjugation**, *P* stands for parity inversion. The charge conjugation, *C*, is the operation to flip the electrical charge of the particle, keeping the other parameters same. For example, if we operate *C* to an electron, the electron changes its electrical charge, and it becomes a positron. So by a C operation, the particle changes to its antiparticle, or vice verse. The CP operation means to operate both C and P. And the CP symmetry, or CP invariance, means that the particle does not change its behavior by flipping CP.

So why do we take CP instead of simple C? We may think that C can change a particle to its antiparticle, period. This is true as long as the particle is only interacting electromagnetically. Inversion of electric charge converts a particle to its antiparticle perfectly if only electromagnetic interaction is concerned. However, if we consider the weak interactions, the inversion of electric charge is not enough to convert a particle to its antiparticle. As we mentioned earlier, the neutrinos from β decays are left handed. Their spin directions are opposite to their motions. What about the helicities of antineutrinos? Though neutrinos do not have electrical charge, we can assume that the neutrinos will change to antineutrinos by a C operation. We can also assume that antineutrinos are emitted by the β^+ decays, with accompanying the positrons. By studying the helicities of the positrons, it turned out that antineutrinos from β^+ decays are right handed. So in the β decay processes, the particle to antiparticle correspondence is that the left handed neutrino is to the right handed antineutrino, or vice verse. And as we mentioned earlier, P operation flips the handedness: left hand to right hand, or vice verse. So to convert the particle to its antiparticle, we have to operate both C and P if we take into account the weak interactions. That is why we study the CP symmetry. Because we already know that both the P symmetry and the C symmetry are violated in the weak interactions.

left hand neutrino ↔ right hand antineutrino

In addition, the weak interaction theory of Feynman and Gell-Mann are also invariant under the CP inversion. So things seemed to be happy with the CP symmetry.

But in 1964, the CP symmetry was also found out to be broken in the experiment, done by J. Cronin, V. Fitch, et al. They studied the decays of neutral kaons, K^0. The neutral kaons were produced by strong interactions. When neutral kaons were made, there were two types: K^0 and $\bar{K}^0$(anti-K^0). Since the kaon is a meson, it is made of a quark and an antiquark. In the case of K^0, the contents are a d quark and an anti-s quark. For $\bar{K}^0$, they are an s quark and an anti-d quark. By way of weak interactions, K^0 and $\bar{K}^0$ can change to each other. This is possible since they are composed particles. True elementary particles can not change to their antiparticles. In addition, quarks can change their types by weak interactions. The compositeness and weak interaction together enable K^0 to change to $\bar{K}^0$, or vice verse, as shown in the figure below.

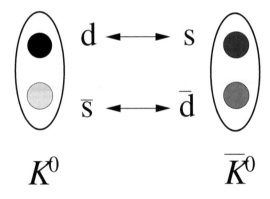

In the above, doubly change of $d \leftrightarrow s$ and anti-$s \leftrightarrow$ anti-d converts $K^0 \leftrightarrow \bar{K}^0$.

When the K^0 or $\bar{K}^0$ decays, they decay by weak interactions. Its decay to leptons, like electron, muon or neutrino, is strongly suppressed because the annihilation of the neutral kaon requires the transition between s quark and d quark, which is strongly suppressed in the weak interaction. As for its decay to hadrons, the neutral kaon is the second lightest meson after pion. Hence, it decays to pions, either two pions or three pions. Its decay to four pions is not allowed because four pions are heavier than the neutral kaon. As for the two pion decays, the two-pion system has a distinct CP value. For a C operation,

$$C : \pi^+\pi^- \rightarrow \pi^-\pi^+$$

or

$$C : \pi^0 \pi^0 \rightarrow \pi^0 \pi^0$$

So after the C operation, the two-pion system returns to itself.

The parity of the pion is known as odd. So a P operation to the two-pion system is

$$P : P = (-1)^2 = +1.$$

It is known that the two pions from the neutral kaons have no orbital angular momentum. So orbital motion has no effect to CP. As a result, the two-pion system has a distinct $CP = +1$ (even).

As for the three pion decays, the three-pion system also has a distinct CP. For a C operation,

$$C : \pi^+ \pi^- \pi^0 \rightarrow \pi^- \pi^+ \pi^0 \rightarrow P(\pi^+ \pi^-)\pi^0 \rightarrow \pi^+ \pi^- \pi^0$$

where $\pi^- \pi^+$ was swapped by a parity operation on the two pion system. For three neutral pions,

$$C : \pi^0 \pi^0 \pi^0 \rightarrow \pi^0 \pi^0 \pi^0$$

As a result, $C = +1$ in the both cases.

For a P operation,

$$P : P = (-1)^3 = -1.$$

As a result, the three-pion system has a distinct $CP = -1$ (odd).

In the weak interactions, CP symmetry is considered to hold, or as least it holds approximately. If the CP value does not change in the decay process, the neutral kaon that decays to two pions has to have $CP = +1$ (even), while the neutral kaon that decays to three pions has to have $CP = -1$ (odd).

Does K^0 or $\bar{K}^0$ have a distinct CP? It turned out that neither K^0 nor $\bar{K}^0$ has a distinct CP because a C operation changes it as

$$C : K^0 \rightarrow \bar{K}^0.$$

In order the parent kaons to have a distinct CP, they have to be mixed: either

$$K^0(50\%) + \bar{K}^0(50\%)$$

or

$$K^0(50\%) - \bar{K}^0(50\%).$$

In the above mixtures, two simultaneous transitions by a C operation, i.e. $K^0 \rightarrow \bar{K}^0$ and $\bar{K}^0 \rightarrow K^0$, cancel out, and they have distinct CPs.

The decay lifetimes of two pion decay and three pion decay are quite different. Since the total mass of three pions is only slightly lighter than the neutral kaon, the three-pion system has extremely limited combinations of the decayed pions' momenta. As a result, the decay is suppressed. Hence the lifetime is long. On the other hand, the total mass of two pions is much lighter than the neutral kaon, thus there is little such restriction. Hence the decay is not suppressed and the lifetime is short.

Because of this difference in lifetimes, the neutral kaons that decay to three pions are called K_L^0 (L stands for long), and the neutral kaons that decay to two pions are called K_S^0 (S stands for short).

In the Cronin el al's experiment, they produced neutral kaons by hadronic collisions via strong interactions. Then, after a long time from the productions, long enough for all the K_S^0s had decayed, they measured the decay mode. Since there were only K_L^0s left, all the decays should be three pion decays from the K_L^0s. However, they observed two pion decays at a rate of one in every 500 decays. This means that K_L^0 (CP odd) decayed to two pions (CP even). Hence, the CP symmetry is broken.

10.8 Kobayashi-Maskawa theory

Decays of hadrons with s quarks, so called strange hadrons, are suppressed compared to the hadrons without s quarks, so called non-strange hadrons. To explain this, using the Fermi theory, strange hadrons and non-strange hadrons have to have different constants in the theory. In addition, these constants are also different from the constant for muon decays.

If we write those constants as G_s for strange hadrons, G_n for non-strange hadrons, and G_F for muons, N. Cabibbo found an interesting relation between them as

$$G_s^2 + G_n^2 = G_F^2.$$

If we divide both sides of the above equation by G_F^2, we get

$$\left(\frac{G_s}{G_F}\right)^2 + \left(\frac{G_n}{G_F}\right)^2 = 1.$$

This has a similarity with a trigonometry equation

$$\sin^2\theta + \cos^2\theta = 1.$$

So he used trigonometric functions to express the constants as

$$G_s = G_F \sin\theta_C$$
$$G_n = G_F \cos\theta_C$$

where we call G_F as the **Fermi coupling constant**, θ_C as the **Cabibbo angle**. Their relation is graphically drawn below.

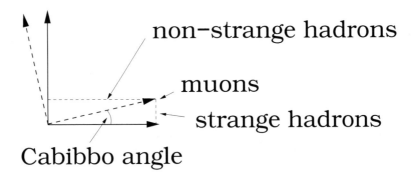

In 1973, Kobayashi and Maskawa found that if the Cabibbo's idea is extended to six quarks, *CP* violation can be explained. In 1973, we only knew three quarks, *u*, *d* and *s*. In such a circumstance, assuming the existence of six quarks was outrageous. But they did it. With the six quarks, quarks can be written by three pairs as shown in the figure (a) below.

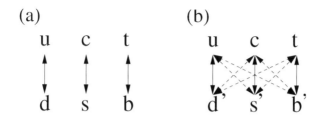

By weak interactions, those quarks can change as $u \leftrightarrow d$, $c \leftrightarrow s$, $t \leftrightarrow b$, as shown in (a). As was known at that time, there is a cross line between $u \leftrightarrow s$, in addition to those vertical lines in the figure. The interaction by this cross line was described by the Cabibbo rotation. Kobayashi and Maskawa extended this cross line to six quarks as shown in (b). Since these transitions are described by the quantum mechanics, they should be expressed by a unitary matrix with three columns and three rows (3×3). This matrix is called the **Cabibbo-Kobayashi-Maskawa (CKM) Matrix**.

Maskawa at that time noticed that 3×3 unitary matrices, in general, have an imaginary number in their components. What does this mean? In the Fermi theory with the CKM matrix, it is known that if CP is inverted, the matrix will change to its complex conjugate. It is rather cumbersome to show this, so we skip to prove this in this book. The complex conjugate is a complex number with the sign of its imaginary part flipped from the original number. The figure below shows a complex number and its conjugate.

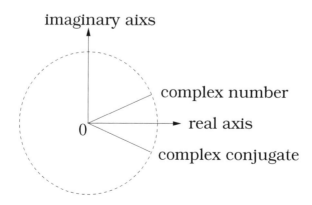

So if there is a complex number in the CKM matrix, which is generally true, interactions described by the Fermi theory will be different after the CP inversion. Since they are different before and after the CP inversion, this causes the CP violation.

As shown in the above figure, a complex number and its conjugate have phases with opposite sign. From this, when a CKM matrix changes to its complex conjugate, there will be a difference in the phases of the oscillations, or waves. If the difference in phase is only in a single wave, we can not detect the difference since in quantum mechanics simple difference in phase has no effect on the phenomena. However, if the wave interferes with other waves, there can be differences in the amplitude of the combined wave. Hence, we have a chance to detect the differences.

A. I. Sanda, I. I. Bigi and A. B. Carter found a clever method to measure this effect, caused by the complex number nature of the CKM matrix, using B mesons. We will explain their method, though it is somewhat complicated.

First, what is B meson? B meson is a meson with a b quark or an anti-b quark in its content. This naming is similar to the strange meson that has either an s quark or an anti-s quark. In the Sanda et al's method, neutral B mesons are used. The neutral B mesons are either B^0 or $\bar{B}^0$. B^0 consists of a d quark and an anti-b quark while $\bar{B}^0$ consists of a b quark and an anti-d quark. Those neutral B mesons can change each other, similarly to the case of the neutral K mesons as shown below.

$$B^0 \qquad\qquad \bar{B}^0$$

In the above figure, doubly change of $d \leftrightarrow b$ and anti-$b \leftrightarrow$ anti-d converts $B^0 \leftrightarrow \bar{B}^0$.

A difference between the B^0 case and the K^0 case is the difference in the quark generations. In the B^0 case, the change is between the first generation quark, d, and the third generation quark, b, while in the K^0 case, the change is between the first generation quark, d, and the second generation quark, s. The change between the first generation and the second generation results no significant phase change. On the other hand, the change between the first generation and the third generation results phase change. The reason of this difference is not known to us. We can just say that the nature is made that way. The transition between the first generation and the third generation is special in that sense by an unknown reason. As a result, a transition: $B^0 \leftrightarrow \bar{B}^0$ causes a phase shift.

The neutral B meson has an another useful character. Mixtures of B^0 and $\bar{B}^0$ can form a heavy B^0 and a light B^0. When the neutral B meson decays, it decays either as a heavy B^0 or as a light B^0. In addition, the difference in the mass between the heavy B^0 and the light B^0 is extremely small: a trillionth, 10^{-12}, of their masses. As a result, the difference in their wavelengths, or frequencies, is a trillionth of their wavelength. This tiny difference is crucial to the measurement.

Sanda et al selected a reaction that the neutral B meson decays into $J/\psi K^0$. The reason, why this reaction is selected, is that both B^0 and $\bar{B}^0$ can decay into $J/\psi K^0$. In addition, J/ψ is a meson which can be easily identified by its decay into a pair of leptons, namely electrons or muons.

$$\begin{aligned} B^0 &\rightarrow J/\psi + K^0 \\ \bar{B}^0 &\rightarrow J/\psi + K^0 \end{aligned}$$

Because of this characteristics, there are two ways for the B^0 to decay to $J/\psi K^0$: (a) B^0 changes to $\bar{B}^0$ first, then the $\bar{B}^0$ decays to $J/\psi K^0$, (b) B^0 directly decays to $J/\psi K^0$. In addition, we can swap B^0 and

$\bar{B}^0$ in the above reactions.

$$B^0 \rightarrow \bar{B}^0 \rightarrow J/\psi + K^0$$

or

$$\bar{B}^0 \rightarrow B^0 \rightarrow J/\psi + K^0$$

It is also possible that B^0 changes twice before decaying to $J/\psi K^0$. But we can ignore such multiple change cases because B^0 mostly decays to $J/\psi K^0$ before it changes back from its antiparticle.

When the particle was B^0 at the beginning, there are two ways to decay into $J/\psi K^0$

$$\begin{aligned} B^0 &\rightarrow & J/\psi + K^0 \\ B^0 &\rightarrow \bar{B}^0 \rightarrow & J/\psi + K^0. \end{aligned}$$

For these reactions, there is no way to distinguish which reaction in the above that the B^0 actually took. However, when B^0 changes to $\bar{B}^0$, the phase of its wave changes. So there is a difference in phase between the direct ($B^0 \rightarrow J/\psi + K^0$) and the indirect ($B^0 \rightarrow \bar{B}^0 \rightarrow J/\psi + K^0$) decays. As a result, those two waves interfere each other. Thus the combined wave will be modulated.

But this modulation can not be measured in the present technology because the wavelength is extremely short, 10^{-15}m. However, we can utilize the tiny mass difference between the heavy and the light B^0 mesons. When B^0 decays, it decays either through its heavy state or its light state. The difference in the mass means the difference in wavelength. In this case, the fractional difference is 10^{-12}. If two waves with slightly different wavelengths interface, they will **beat**.

The beat's period will be given by the original wavelength, multiplied by the inverse of the fractional difference of the wavelengths.

$$(\text{wavelength of beat}) = \frac{(\text{original wavelength})}{(\text{fractional difference of wavelengths})}$$

Thus, the original wavelength is multiplied by 10^{12} in this case. As a result, the scale of the modulation pattern becomes $10^{-15} \times 10^{12} = 10^{-3}$m. That is 1mm. By the precise calculations, it is shorter than 1mm, but it is still within our measuring capabilities.

The interference and beats of those waves of four different decay paths are illustrated below.

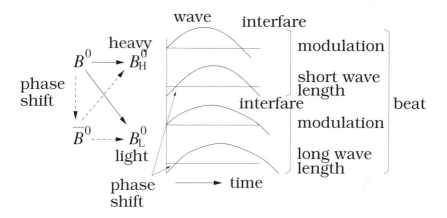

Next, we flip B^0 and $\bar{B}^0$ in the above discussion. When the initial particle is $\bar{B}^0$, the story is the same as the B^0 case, except its phase. Since $\bar{B}^0$ is the complex conjugate of B^0, the sign of its phase is opposite. As a result, the signs of interference and modulation also become opposite, thus the observed modulation pattern is flipped from the B^0 case.

As a summary, the Kobayashi-Maskawa theory predicts that, when B^0 and $\bar{B}^0$ decay into $J/\psi K^0$, their decay frequency patterns, as a function of time, are opposite, as illustrated below.

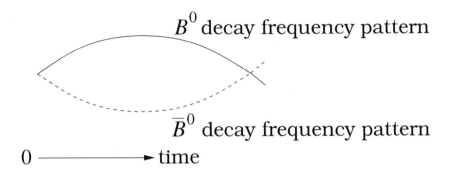

B^0 decay frequency pattern

$\overline{B}^0$ decay frequency pattern

0 ──────→ time

In the above figure, the time scale is in the order of pico seconds (ps).

Using this very clever method, very difficult experiments were done in the United State and in Japan, about the same time around 2000. The results of the both experiments were consistent with the prediction from the Kobayashi-Maskawa theory.

10.9   Spontaneous symmetry breaking

Physicists think that masses of elementary particles should be zero with few exceptions. But the real particles are mostly massive. Spontaneous symmetry breaking can solve this problem. Why the masses should be zero?

We start from the particles that mediate forces, i.e. gauge particles, to explain the massless nature of the elementary particles. First, photons are really massless. We had already explained this. Gluons are also massless as their original forms. However, the gluons get their masses from strong interactions. Strong interaction energy, accompanying to the gluon, produces a mass of the gluon. We will come back to this later.

As for the particles that mediate weak interactions, they seem to have masses. The reason, why the weak interaction is weak, is due to its extremely short ranginess. Its range is one hundredth of the radius of a nucleus. As we explained earlier, range is inversely proportional to the mass. So in the weak interactions, the mass of the mediating particle should be extremely heavy.

Let us review the force mediating particles. In the electromagnetic forces, the electric charge is constant. Because of this rigidness of the charge, the canonical conjugate of the charge in the Heisenberg uncertainty principle, i.e. gauge function, is arbitrary, or gauge invariant. By extending this gauge invariance to be satisfied at any space point, electromagnetic field is required. And to mediate the electromagnetic field, photons are introduced.

As for the strong force, it is similar to the electromagnetic force. The color charge invariance requires the strong interaction field, or gluon field. Three types of color charge, instead of single type of electric charge, make the underlying symmetry more complex. This complexity causes the gluon-to-gluon interactions.

So what is the underlying mechanism of weak interactions? Are there any charges that are constant? It turned out that there are such charges in the weak interactions too. We call such charge as **weak charge**. Actually weak interactions and electromagnetic interactions are entangled. This was first noticed by S. Glashow in the early 1960s. The mediating particle, which is required by the series of arguments: i.e. constant charge $\rightarrow$ Heisenberg uncertainty principle $\rightarrow$ gauge invariance $\rightarrow$ gauge field $\rightarrow$ gauge particle, has to be massless. If it is not massless, the above series of argument does not hold, while the short ranginess of the weak interaction requires a heavy mass to that particle.

To solve this dilemma, a clever method to give a mass to the massless particle was discovered, which is spontaneous symmetry breaking.

The **spontaneous symmetry breaking** was first introduced to explain the superconductivity, found in some types of ultra cold metal. To make this mechanism work, "base material" has to exist. Then the **phase transition** in the base material is necessary to enable this mechanism. Here the phase transition means the change of material's state. For example, liquid water changes to solid ice at 0°C. This is an example of the phase transition.

Let's think about ferromagnetic material such as iron. The ferromagnetic material can become a magnet. The reason for this magnetic characteristics originally comes from the electron's feature. Electrons are tiny magnets. The electron as a magnet gives magnetism to the iron atom. If the directions of the iron atoms in the material align in the same direction, the material becomes ferromagnetic. Ferromagnetic materials loose their magnetism at high temperature. This is because at high temperature the directions of the iron atoms get randomized as shown below.

high temperature   low temperature

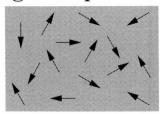 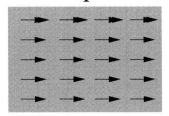

random directions   aligned directions

This transition, changing from the randomly pointing state at high temperature to the aligned (or ordered) state at low temperature, is also a phase transition. At high temperature, this material is symmetric under the change of direction, i.e. rotational symmetry. At low temperature, it looses the rotational symmetry. This phenomenon, loosing the rotational symmetry, is called **spontaneous symmetry breaking**.

When spontaneous breaking of symmetry takes place and the directions of the iron atoms get aligned, the material becomes to have a special direction. Then it is possible to oscillate with respect to this direction, as shown in the figure below.

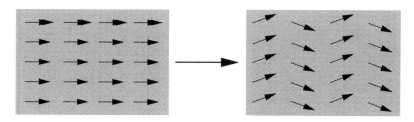

aligned wave

From the wave-particle duality of quantum mechanics, this oscillation is also considered as a particle. Since the oscillation extends to the entire material, uncertainty in the location is huge, or infinite. This, with the Heisenberg uncertainty principle, indicates that this particle is massless. We call this particle **Nambu-Goldstone particle**.

As we have seen in the above case, when a continuous symmetry is broken spontaneously, a massless Nambu-Goldstone particle appears in the material. What is the continuous symmetry? The system is said to have a continuous symmetry, if the system does not change, when a symmetry variable is changed by an arbitrary amount. Rotational symmetry is a continuous symmetry, because it keeps the system same by rotating the system with any angle. The opposite to the continuous symmetry is the discrete symmetry. In the discrete symmetry, the change is digital, like switch on/off.

Y. Nambu is the first physicist who introduced the spontaneous symmetry breaking to the particle physics. In the early 1960s, Nambu applied this mechanism to neutron's β decays. Neutron's β decays can be treated by two ways; (a) the neutron directly decays to a proton, an electron and an anti-electron neutrino, (b) the neutron is always accompanying pions in its surrounding cloud and one of the pions decays to an electron and an anti-electron neutrino. Those two cases are shown below.

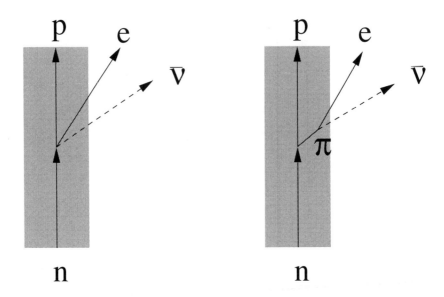

Protons have electric charges. The strength of this charge does not change even the proton is surrounded with pions. Some of the pions have electric charges by themselves, but because of the conservation law of electric charge, the total charges is constant. The electric charge is conserved under the influence of strong interactions.

The protons also have "weak charges." (This weak charge is different from the weak charge in the Weinberg-Salam theory, described later.) If we assume that this weak charge is also conserved under the influence of strong interaction, Nambu derived the following relation.

$$Aq^2 + B = 0 \ \leftarrow \ \text{if weak charge conserves.}$$

Here A and B describe the effects of the surrounding pion cloud around the proton. The q^2 is a quantity similar to the square of mass.

In this argument, the base material for the spontaneous symmetry breaking is the pion cloud, induced by the strong interactions of proton and pions. If the weak charge looses it conservation character because of the existence of this material, or in the other words, if the weak charge conservation law is spontaneously broken in the pion cloud, it can be shown

$$B \neq 0.$$

Actually, experimental measurements show that the weak charge's conservation law is slightly broken.

The above two equations require

$$Aq^2 \neq 0.$$

Because this relation has to hold even for $q^2 = 0$, A has to be written as

$$A = \frac{C}{q^2}.$$

On the other hand, if we treat this decay process through the pion decays, as shown in the right figure above, it can be derived as

$$A = \frac{C}{q^2 - m^2}.$$

Comparing the above two equations, the two equations coincide when the pion mass, $m = 0$. From this argument, pions can be considered as a massless Nambu-Goldstone particle in the hadronic cloud. In reality, since the conservation law of weak charge is an approximate law from the beginning, pions do have masses. But this can explain why the pions are unusually light compared to the other hadrons.

10.10   Higgs mechanism

By applying the idea of spontaneous symmetry breaking further into
the elementary particle physics, P. Higgs, F. Englert and R. Brout
discovered a method to give a mass to the massless gauge particle.
This method is called **Higgs mechanism**.

First, as the base material for phase transitions, we introduce **Higgs
field**. The Higgs field has a characteristics similar to the ferromagnetic
material. When the Higgs field has a phase transition, the internal
symmetry in the Higgs field, a 2D rotation, is broken spontaneously.
This is similar to the case of the ferromagnetic material for which the
rotational symmetry at high temperature is broken at low temperature.
For the Higgs case, we think that the symmetry was broken right after
the Big Bang. The mechanism of symmetry breaking is shown in
the figure below. The Higgs field has two components, h1 and h2.
Originally the mixing ratio of those two components is arbitrary. Any
mixing ratio can be chosen, i.e. 2D rotational symmetry in this h1 and
h2 plane. But at a certain time after the Big Bang, the Universe was
cooled down and this ratio was fixed.

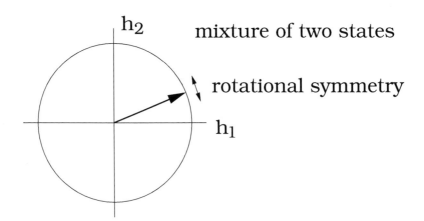

Once the symmetry is fixed, there can be an oscillation around the
fixed point. In the Higgs field case, it is an oscillation in the mixing

ratio, as illustrated below.

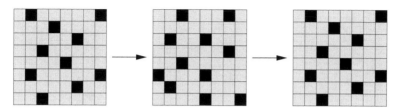

Oscillation of Mixing Ratio

Since the Higgs field extends to the entire universe, the regularity appeared in the whole universe by this phase transition. Then it became possible to deviate from this perfect regularity. This deviations can oscillate. Since it extends to the whole universe, its range is infinite, thus it has to be massless, a Nambu-Goldstone (NG) particle.

Now we remember that force mediating particles, i.e. gauge particles, are also massless, thus they extend to the whole universe too.

Here we assume that the gauge field for weak interactions interacts with the Higgs field. Then, it is possible that the cost zero oscillations of the Higgs field, i.e. the NG particles, can be cancelled out by the associated oscillations of this gauge field. Originally the gauge field for weak interactions worked on the matter field, but the same gauge field also happens to absorb the oscillations of the Higgs field. Physicists call this phenomena as "gauge particles eat NG particles," as illustrated below.

eats   confined

$m = 0$
NG particle

gauge particle
$m = 0$ $m \neq 0$

Since the Higgs field is the base material for the spontaneous symmetry breaking phenomena, it has to have non-zero constant base-energy. Because the gauge field interacts with the Higgs field, the gauge field also interacts with this constant energy. When the oscillations of the Higgs field are absorbed by the gauge field, gauge particles also absorb this constant energy at the same time. This constant energy, absorbed by the gauge particle, behaves as the mass of the gauge particle. As a result, the NG particle disappears and the gauge particle obtains mass. This is called **Higgs mechanism**.

Since the above description of the Higgs mechanism is somewhat complicated, let's explain the Higgs mechanism using a water vessel model. The figure, below, shows the model: the Higgs field corresponds to the water in the vessel and the gauge field corresponds to the vibrating plate. In this model, the NG particle corresponds to the horizontal oscillation of the water (= Higgs field) in the vessel as illustrated on the left side in the figure. By attaching the vessel on the vibrating plate (= gauge field), the vibrating plate can make the vessel oscillate. This oscillation can stop the horizontal oscillation (= NG particle) of the Higgs field as illustrated on the right side in the figure. As a result, the NG particle disappears. And the weight of the water (= Higgs field) presses the vibrating plate (= gauge field) which makes the gauge particle massive. The above is the description of Higgs mechanism by the water vessel model.

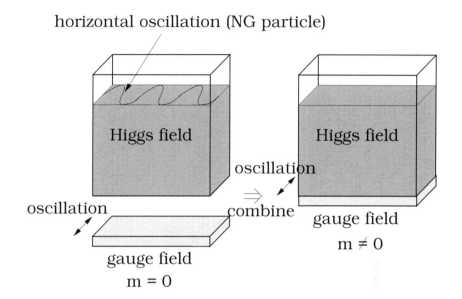

horizontal oscillation (NG particle)

10.11 Weinberg-Salam theory

Force mediating particles for the weak interaction are massless as their original form, but they acquire mass through the Higgs mechanism. As for the matter particles, they also have to be massless.

As we have learned, parity is violated in weak interactions. This was due to the fact that the weak interactions only work on the left-handed matter particles. There is no weak force works on the right-handed matter particles.

Handedness of the particles, however, may not stay constant for all the observers. As you remember, the right-handed means that the spin direction is the same as the moving direction, and the left-handed means that the two directions are opposite. Here, the spin direction is always the same independent of the observers. The moving direction, however, may not be the same for all the observers as shown in the figure below.

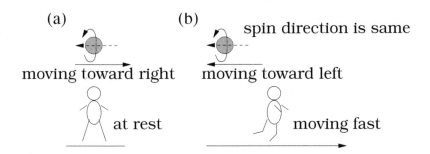

In the above figure (a), the particle is moving toward right, observed by the person at rest. The same particle is observed as moving toward left by the another person who is moving toward right with the velocity faster than the particle, as shown in (b). As a result, the same particle is observed as left-handed by the person at rest, but right-handed by the another person who is moving faster than the particle.

Since weak interactions only work on the left-handed particles, they can work on the particle, seen by the observer at rest, but for the another observer, they can not work, though the particle is the same. This is a conflict.

To avoid this conflict, moving directions should be kept same for all the observers. This is possible if the particle is always moving at the light speed. Since no observer can move faster than the light speed, no one can pass the particle with the light speed. Thus, the particle's moving direction will be kept same. To enable the particle to move with the light speed, the particle has to be massless. To accelerate a massive particle to the light speed, infinite energy is necessary, which is impossible. So to avoid the conflict in the weak interactions, all the matter particles have to be massless.

However, the actual particles have masses. So the massless particles have to acquire masses by some mechanism. Weinberg proposed such a mechanism. He used the Higgs field for this purpose too.

If a matter particle, moving at the light speed, interacts with the Higgs field, it is scattered by the Higgs field. By those scatterings, the matter particle moves randomly as a Brownian motion in a microscopic scale, as shown in the figure below.

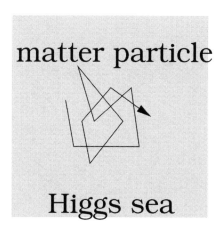

As illustrated in the above figure, if a matter particle, moving with the speed of light, is in the Higgs field, or in the "Higgs sea," the particle is scattered randomly by the sea water. This effect restricts the location of the matter particle. As you remember, the range of spread is inversely proportional to the mass. In this case, the restriction on the location gives mass to the matter particle. As a result, the massless matter particle acquires the mass. In addition, apparent speed of the particle will be slower than the light speed because of these scatterings.

The scattering frequency of the matter particle with the Higgs field depends on the particle's kind. As shown at the left side in the figure below, the particle, that scatters with low frequencies in the Higgs sea, has a large occupying region. Thus such a particle is light. On the other hand, as shown at the right side in the figure, the particle that scatters with high frequencies has a small occupying region. Thus such a particle is heavy. As we have seen, the scattering frequency determines the particle's mass. Also the higher scattering frequency is caused by a stronger interaction strength between the matter particle and the Higgs field. The interaction between the matter particle and the Higgs field is called as **Yukawa interaction**. So, if the strength of Yukawa interaction is large, the matter particle will be heavy, if the strength is small, the matter particle will be light. The Yukawa interaction does not belong to any of the electromagnetic, or strong, or weak interactions.

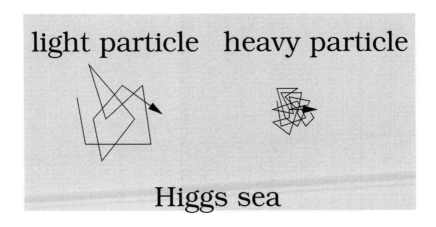

As a summary up to here, the Higgs field provides mass to the force mediating particles by the Higgs mechanism. In addition, the Higgs field is re-used to provide mass to the matter particles by the Yukawa interaction.

As for the conservation of weak charge, it entangles with electric charge, as we had mentioned at the beginning of this chapter. Their corresponding gauge functions give gauge particles, W^+, W^-, Z^0 as well as the photon. Though photons are free from the Higgs mechanism, thus, massless, other three are massive by the Higgs mechanism. In addition, $W^{\pm}$ and Z^0 are based on the non-abelian group, thus they interact with themselves. They also show the characteristics of asymptotic freedom. Contrary to the strong interaction, however, those force mediating particles can be observed as free particles. The above are the outline of the Weinberg-Salam theory. The theory was published in 1967. The theory predicted so called "neutral current," which had not been observed at that time. The neutral current is the phenomenon that's the weak interaction mediated by the neutral particle, Z^0. In the early 1970s, the neutral current was experimentally found in neutrino-electron collisions. Finally, in the early 1980s, $W^{\pm}$ and Z^0 were discovered by C. Rubia et al at CERN, which confirmed the correctness of the theory.

10.12 Renormalization

Shortly after the discovery of the neutral current, M. Veltman and G. 't Hooft proved that the Weinberg-Salam theory is **renormalizable**. We will explain what is renormalizable.

The figure, shown below, illustrates "water," looking at the various scales.

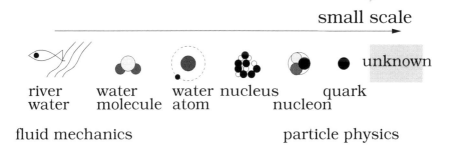

When we look at water in our daily life, it can be seen as the water in the river, for example. If we close up the water, we can find that the water is made of water molecules. The water molecule consists of one oxygen atom and two hydrogen atoms. The oxygen atom is made of a nucleus and electrons. The nucleus is made of nucleons. And finally, the nucleon is made of quarks. At the present knowledge, the electron and the quarks are the smallest objects. We consider them as point particles and they can not be divided into smaller components.

In the each level of the scale, we have a corresponding physics. For the river water, it is fluid dynamics. For the water molecule, it is molecular physics, or chemistry. For the water atom, it is atomic physics. For the oxygen nucleus, it is nuclear physics. For the nucleon, electron and quarks, it is elementary particle physics.

Using the fluid dynamics, water in the river can be described. When we use the fluid dynamics, we do not need the knowledge of oxygen nucleus or its quark contents. All the smaller scale phenomena are included in the parameters in the fluid dynamics: density of water, viscosity of water etc. Those parameters are usually obtained by experiments. They also could be derived by using the molecular physics, which is one level smaller than the fluid dynamics. But we do no need the knowledge of quarks, for example, to derive those parameters.

As seen above, a phenomenon in a certain scale can be described by the physics, corresponding to that scale. The effects from the smaller scale physics are put together into the parameters, used in the physics of that level. Those parameters can be derived by either through the experiments or using the knowledge of the physics, just one level smaller than that level. Physics should be made up of such layered structures. This is the idea of **renormalization.**

If a physics is not made up of the above way, it may need the knowledge of quarks to describe the behavior of the water in the river. Such physics is said to be not renormalizable. A renormalizable physics does not require the knowledge of physics of smaller scales. If a physics is not renormalizable, it is impossible to use the physics. So a valuable physics has to be renormalizable.

Now we are back to the particle physics. We do not know the physics smaller than the physics for electrons or quarks, i.e. elementary particles. We treat those elementary particles as points. But there are reasons to believe that they are not points.

If an electron is a point, for example, the radius of the electron is zero. The general relativity, however, predicts a Schwarzschild radius for the mass of the electron. The Schwarzschild radius is a distance that the gravity gets too strong for anything, even a photon, to escape, as shown in the figure below. Since the Schwarzschild radius is finite, the point electron can be smaller than the Schwarzschild radius. We can imagine this without knowing the general relativity. In the Newton's gravity law, the strength of gravity is inversely proportional to the square of the distance. Thus, if the distance is zero, the strength will be infinite. If this is true, the electron becomes a **blackhole**. Since electrons are not blackholes, the size of the electron has to be finite, or the theories of gravity have to be modified in the extremely small scale.

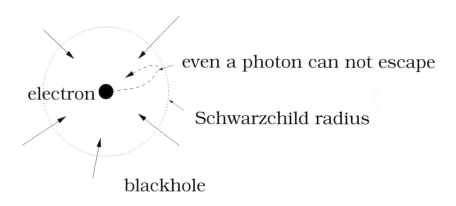

As shown in this example, the extremely small world has a full of mystery. Our current knowledge does not work at such a small world. But the elementary particle physics has to work without such knowledge. Otherwise it can not be used. Thus the effects from the extremely small world have to be put together into the parameters used in the particle physics, i.e. the theory has to be renormalizable. In this case, the effects should be put into the electron's mass, charge etc.

In the elementary particle physics, there are extra requirements about the renormalizability. That is a small scale effect of the theory itself. For example, an electron will emit a virtual photon, which will be converted to a virtual electron-positron pair, and so on. Within the Heisenberg uncertainty principle, there can be infinite numbers of virtual processes. The effects from those virtual processes are called **radiative corrections**. Veltman and 't Hooft proved that such radiative corrections can be renormalized in the Weinberg-Salam theory. So the Weinberg-Salam theory passed the requirement for usable theories.

10.13   Higgs particle

The Higgs field can give masses to both the massless gauge particles: $W^{\pm}$, Z^0 and the massless matter particles: electrons, quarks, etc. The Higgs field also creates an interesting particle: **Higgs particle**.

As we had shown in the description of the Higgs mechanism, two components of the Higgs field, h1 and h2, form a 2D plane as shown in the figure below. The Higgs field can oscillate along the radial direction at the fixed mixing ratio, in addition to rotate round the center.

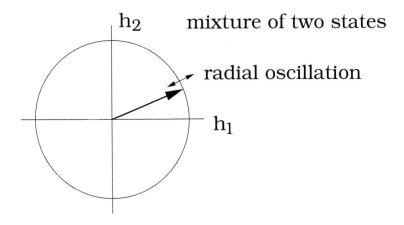

Since there is no symmetry in the radial direction from the beginning, this oscillation does not produce NG particles. Instead, it produces an "ordinary particle," i.e. the Higgs particles. It is an oscillation of the size of a cluster (nugget) in the Higgs field, as illustrated below.

oscillation of cluster size

Since Higgs field is a kind of material that fills the vacuum in the entire universe, its characteristics has to be the same as vacuum. And since the Higgs particles are its oscillations, Higgs particles also inherit the characteristics of vacuum. Thus, Higgs particles can not have internal features such as spin, contrary to the electrons that have spins or the photons that have polarization. Such particles, with no special characters except its mass, are called **scalar particle**. The Higgs particles are scalar particles.

The Higgs particle can interact with itself, making a nugget. By becoming a nugget, its location is limited, thus it gets mass. So Higgs particles are massive particles. This interaction is also outside the classification of electromagnetic, strong or weak interactions.

Higgs particles were predicted by Higgs in 1964. However, they had not been discovered until 2012. It took about 50 years before they were found by two experiments: ATLAS and CMS, at CERN.

The reasons why it took such a long time to discover the Higgs particle are

- Higgs particle's mass could not be predicted from the theory by using the known data.

- Higgs particle's mass was very heavy. It is about 130 times the mass of a proton. This requires a very high energy accelerator to produce Higgs particles.

- Higgs particles interacts very weakly with other particles. Thus Higgs particles are rarely produced. This made it impossible to find the Higgs particles in the cosmic-rays.

- Higgs particles decay to other particles with extremely short time. Its lifetime is 10^{-21}s. This also disabled to find the Higgs particles in the cosmic-rays.

By the discovery of Higgs particles, Higgs mechanism is verified. This is astonishing. Though the Higgs mechanism is a very clever method to put a mass to the massless gauge particle, it has a scent of artificial flavor. Many physicists thought it too artificial to be real. However, the nature turned out to be quite artificial, more than many physicists thought.

10.14   Origins of mass

It is natural to assume that all elementary particles are massless as their original forms. Exceptions could be neutrinos. To make the story simple, we assume that all particles are originally massless. Real particles, however, have masses, except photons. The mechanisms that provide masses to these massless particles are summarized in the following list.

particle kind	subdivision	original mass	real mass	mechanism to obtain mass
matter particle	quark	0	$\neq 0$	Yukawa + strong interaction
	lepton	0	$\neq 0$	Yukawa interaction
gauge particle	photon	0	0	
	$W^{\pm}, Z^0$	0	$\neq 0$	Higgs mechanism
	gluon	0	$\neq 0$	strong interaction
Higgs particle	Higgs particle	?	$\neq 0$	interaction with itself

Let's think of the origin of ourselves as human beings. Does our weight originate from the Higgs field? The answer to this question is no in a first approximation. Although the protons, neutrons and electrons, which build our body, are matter particles, thus they get their masses by the Yukawa interactions, the contributions from the Yukawa interactions are only a few percents of our total weight. The majority of our weight comes from the strong interactions. Quarks and gluons have large amount of energies by their strong interactions. These energies contribute to our weight. So our weights are mostly the strong interaction energies.

As for the origins of the neutrinos' masses and the Higgs particles' masses, physicists are still working on them.

Index

Epilogue

If the contents of a book are diagonalized, the book is easier to read for the readers. The word, diagonalize, means that each subject is independent from each other. It is a technical term used in quantum mechanics. When the contents are diagonalized, the subjects are not entangled, thus the reader can understand each subject without knowing the other subjects. When the contents are not diagonalized, the contents are entangled, thus the reader has to understand all the subjects in order to understand one subject. If we express this non-diagonalized in the other words, we can say that those subjects are non-commutative. Or the subjects interact with the other subjects. This is the way that most physics books are. This book is no exception, contents are not diagonalized. Please read all the chapters except possibly the first few.

To avoid using equations, some analogies are used in the various parts in this book. Analogies may be interesting to the persons who already know the subject. But it may just give another confusion to the persons who do not know the subject. Be careful.

As I was writing this book, I felt it may be easier for the readers, if I use equations to describe the subject. Elementary particle physics is probably easier to understand if equations are used. If you have not read a book with some equations in it, I suggest you to read such books after this book. This book, I believe, should be useful when you read advanced books.

Made in the USA
Columbia, SC
21 June 2017